ABS PROJECTION ALGORITHMS:
Mathematical Techniques for Linear and Nonlinear Equations

MATHEMATICS AND ITS APPLICATIONS

Series Editor: G. M. BELL, Professor of Mathematics,
King's College London (KQC), University of London

NUMERICAL ANALYSIS, STATISTICS AND OPERATIONAL RESEARCH

Editor: B. W. CONOLLY, Emeritus Professor of Mathematics (Operational Research), Queen Mary College, University of London

Mathematics and its applications are now awe-inspiring in their scope, variety and depth. Not only is there rapid growth in pure mathematics and its applications to the traditional fields of the physical sciences, engineering and statistics, but new fields of application are emerging in biology, ecology and social organization. The user of mathematics must assimilate subtle new techniques and also learn to handle the great power of the computer efficiently and economically.

The need for clear, concise and authoritative texts is thus greater than ever and our series will endeavour to supply this need. It aims to be comprehensive and yet flexible. Works surveying recent research will introduce new areas and up-to-date mathematical methods. Undergraduate texts on established topics will stimulate student interest by including applications relevant at the present day. The series will also include selected volumes of lecture notes which will enable certain important topics to be presented earlier than would otherwise be possible.

In all these ways it is hoped to render a valuable service to those who learn, teach, develop and use mathematics.

Mathematics and its Applications

Series Editor: G. M. BELL, Professor of Mathematics, King's College London (KQC), University of London

Series continued at back of book

ABS PROJECTION ALGORITHMS:
Mathematical Techniques for Linear and Nonlinear Equations

JOZSEF ABAFFY, B.Sc., Ph.D.
Department of Mathematics & Computer Science
University of Economics, Budapest, Hungary

EMILIO SPEDICATO, Laurea
Department of Mathematics
University of Bergamo, Italy

ELLIS HORWOOD LIMITED
Publishers · Chichester

Halsted Press: a division of
JOHN WILEY & SONS
New York · Chichester · Brisbane · Toronto

First published in 1989 by
ELLIS HORWOOD LIMITED
Market Cross House, Cooper Street,
Chichester, West Sussex, PO19 1EB, England
The publisher's colophon is reproduced from James Gillison's drawing of the ancient Market Cross, Chichester.

Distributors:

Australia and New Zealand:
JACARANDA WILEY LIMITED
GPO Box 859, Brisbane, Queensland 4001, Australia

Canada:
JOHN WILEY & SONS CANADA LIMITED
22 Worcester Road, Rexdale, Ontario, Canada

Europe and Africa:
JOHN WILEY & SONS LIMITED
Baffins Lane, Chichester, West Sussex, England

North and South America and the rest of the world:
Halsted Press: a division of
JOHN WILEY & SONS
605 Third Avenue, New York, NY 10158, USA

South-East Asia
JOHN WILEY & SONS (SEA) PTE LIMITED
37 Jalan Pemimpin # 05–04
Block B, Union Industrial Building, Singapore 2057

Indian Subcontinent
WILEY EASTERN LIMITED
4835/24 Ansari Road
Daryaganj, New Delhi 110002, India

© 1989 J. Abaffy and E. Spedicato/Ellis Horwood Limited

British Library Cataloguing in Publication Data
Jozsef Abaffy
ABS projection algorithms: mathematical techniques for linear and nonlinear equations.
1. Algorithms
I. Title II. Series III. Spedicato, Emilio
511′.8

Library of Congress data available

ISBN 0–7458–0423–3 (Ellis Horwood Limited)
ISBN 0–470–21507–0 (Halsted Press)

Printed in Great Britain by Unwin Bros., Woking

Table of contents

8 **Table of contents**

Preface

During the autumn of 1981 Abaffy, who was then visiting the University of Bergamo for research on quasi-Newton methods, was asked by Spedicato to deliver a seminar on some recent work he had done in Hungary. He talked about a generalization published by him in the not widely read Hungarian journal *Alkalmazott Matematikai Lapok* of an algorithm proposed by Huang in 1975 in the *Journal of Optimization Theory and Applications*. The seminar of Abaffy was the starting point for an intensive collaboration that in the following years has involved not only us but a number of mathematicians from several countries (Italy, Hungary, UK and China). The outcome of this research is the theory of the ABS methods presented in this monograph.

Our work was pursued initially against many doubts of our colleagues in the mathematical community about its theoretical potential and practical usefulness. Currently the ABS methods are attracting a growing interest, as is shown by the publication of papers outside our group and by the number of PhD students working in this field (one of us met four such students during a recent trip to China). We hope that the material presented in this monograph, based upon more than 50 reports and papers on ABS methods, justifies our opinion not only that the ABS approach is a theoretical tool unifying many algorithms which are scattered in the literature but also that it provides promising and, in some cases, proven effective techniques for computationally better algorithms for a number of problems.

Our work has benefited from discussions and contributions from many people. We emphasize the fundamental contributions of Charles Broyden, formerly at the University of Essex (UK), and of Aurel Galantai, of the University of Gödöllö (Hungary). By introducing the parameter z_i, Broyden enabled us to overcome a difficult impasse; in view of this contribution, Broyden's initial (together with our initials) is found in the name ABS. The high technical skills of Galantai have made possible major inroads in the error analysis and in the convergence analysis for nonlinear systems. We acknowledge the contributions of Lawrence Dixon (Hatfield Polytechnic, UK), Naiyang Deng and Meifang Zhu (Beijing Polytechnical University, China) and Marida Bertocchi (University of Bergamo, Italy). Among the PhD students who have worked on ABS methods, Elena Bodon (INDAM, Rome, Italy), Maria Teresa Vespucci (University of Bergamo, Italy), and Gu Shanhong Zhao Yang (Dalian University of Technology, China) have contributed results which are presented in this monograph.

Among the persons with whom we had stimulating and critical discussions, we thank Professor Josef Stoer (University of Würzburg, West Germany), Professor Per-Åke Wedin (University of Umeå, Sweden) and Gerhardt Zielke (Martin-Luther University, Halle, East Germany). We give particular thanks to Professor Ben Noble, now in retirement in the Lake District (UK), who read most of this monograph while in preparation and gave thorough and thoughtful comments.

Financial support is acknowledged from the Consiglio Nazionale delle Ricerche, particularly from the Programma Professori Visitatori, Comitato Nazionale della Matematica, and from the Science and Engineering Research Council of the UK. Without this financial support the international collaboration which has produced the ABS methods would not have been possible.

This monograph is dedicated by Abaffy to the memory of his parents Erzsébet and Gyula, and by Spedicato to the memory of his parents Angela and Giovanni and his teacher at Liceo Manzoni in Milano, Wanda Ferrari, whose crystal-clear lectures in mathematics led him into this field.

1

Introduction

1.1 SCOPE AND LIMITATIONS OF THE MONOGRAPH

This monograph presents a general class of algorithms for solving systems of linear and nonlinear algebraic equations. These are basic problems in numerical mathematics which can be addressed by a large number of currently known algorithms, many of them available as codes in mathematical software libraries. The contribution of this monograph in a field which has been deeply and thoroughly ploughed by mathematicians, particularly after the advent of the electronic computers, is twofold. The first contribution is to present a unified approach to the majority of the existing algorithms for linear and nonlinear systems. They are embedded in a general class of algorithms, the ABS class (see section 2.3 for an explanation of the name ABS), where they correspond to particular choices of some free parameters. More precisely, in the linear case the ABS class contains essentially all possible algorithms with the following property: they can solve, in exact arithmetic, a linear system starting from an arbitrary point and in a number of iterations no greater than the number of equations (see Chapter 7 for a rigorous formulation of this statement). The majority of the direct-type algorithms proposed in the literature have this property and fall therefore into the ABS formulation (the Quasi-Newton algorithms of Broyden (1965) are a notable exception, since they may require for termination on n equations $2n$ iterations). In the nonlinear case the ABS class contains the Newton method, the methods of Brown and of Brent and some of their generalizations.

The second contribution of the monograph lies in the evidence provided, albeit still limited, that some new algorithms, or new formulations of classical algorithms, are computationally attractive and compete with classical algorithms in terms of accuracy, storage requirement and overhead costs. We give evidence (in particular, see Chapter 10) that the modified Huang algorithm can solve very-ill-conditioned linear systems, or least squares problems, more accurately than some implementations of the LU or the QR algorithms available in commercial libraries. In Chapter 11 we show that some ABS methods can solve certain large structured linear systems more efficiently than classical algorithms can in terms of storage or overhead.

The ABS algorithms are studied in this monograph under the following limitations which will be tacitly assumed in the sequel.

— The computations are performed exactly. We shall give, *passim*, remarks on how round-off affects certain properties. In Chapter 12 we shall provide a general, but by no means exhaustive, error analysis.
— The operations are assumed to be performed in the usual sequential order, disregarding the possibilities offered by vector and parallel computers.
— No attention is paid to the question of optimal data addressing and data storage.

In view of these limitations we do not discuss the question of the optimal software implementation of the algorithms given and we do not present any code or pseudocode implementing ABS algorithms. We are currently involved in the development of such codes which will be documented elsewhere. We also postpone to the future the analysis of efficient implementations of the ABS algorithm on vector and parallel computers, which would be short lived owing to the changing pattern of computer architectures.

1.2 SYNOPSIS

The first 12 chapters of this monograph are concerned with the problem of solving a linear algebraic system of the form

$$Ax = b \ , \tag{1.1}$$

where $A \in R^{m,n}$, $x \in R^n$, $b \in R^m$. In Chapter 9 we consider the case where $m > n$ (overdetermined systems, solved for the minimal Euclidean norm least squares solution); otherwise we assume that $m \leqslant n$ (determined or underdetermined systems). The system (1.1) will also be indicated componentwise by

$$a_i^T x = b^T e_i \ , \qquad i = 1, \ldots, m \ , \tag{1.2}$$

where $e_i \in R^m$ is the ith unit vector and $a_i^T \in R^n$ is the ith row of the coefficient matrix A, i.e.

$$A = \begin{bmatrix} a_1^T \\ . \\ a_m^T \end{bmatrix} \ . \tag{1.3}$$

By introducing the residual vector $r = r(x) \in R^m$ defined by

$$r(x) = Ax - b \ , \tag{1.4}$$

the system (1.1) will also be occasionally indicated by

$$r(x) = 0 \ . \tag{1.5}$$

It will be tacitly assumed that all the rows of A are nonzero. It is not assumed in general that A is full rank. The rank of A, $q = \mathrm{rank}(A)$, will be therefore an integer between 1 and $\min(m, n)$. In order to simplify the statement of various properties, the assumption that A is full rank will often be made.

It is assumed that the reader is familiar with the basic results in linear algebra and numerical analysis, available for instance in the work of Broyden (1975), Noble and Daniel (1977), Graham (1979), Dahlquist and Björck (1974) and Phillips and Cornelius (1986) or for a more advanced treatment in the work of Voyevodin (1983), Golub and Van Loan (1983), Stoer and Bulirsch (1980) and Schendel (1984).

Chapter 2 deals with a derivation of the ABS class in terms of analogies with the quasi–Newton algorithms. The heuristics presented here follows exactly the line of thought that led Abaffy (1979) to obtain a one-parameter generalization of the Huang (1975) algorithm. The connection with the quasi-Newton algorithms is an output of the research of Abaffy in that field (also Broyden and Spedicato have worked in the quasi-Newton field; if the reader is familiar with optimization the connection with this will be felt throughout the book). Of course other approaches are feasible and they are in fact illustrated when we point out the connection with the equivalent algorithms of Stewart (1973) and Broyden (1985).

In Chapter 3 we give a number of fundamental properties of the iterates produced by the basic ABS class.

In Chapter 4 we present various alternative formulations of the ABS class, in terms of different formulas for the search vectors. We discuss to some extent the storage requirement and overhead costs (measured by the number of multiplications, as is usual in the sequential computation framework). Among the most stimulating open problems in the ABS class we identify that of determining optimal formulations in terms of complexity and numerical stability, for sequential, vector and parallel computers.

Chapter 5 deals with the method proposed originally by Huang (1975) in the paper which motivated Abaffy (1979) to develop the ABS class. We study extensively its properties and various alternative formulations. The Huang algorithm plays a special role in the ABS class, related to its ability to compute the minimal Euclidean norm solution of underdetermined linear systems.

In Chapter 6 we consider another fundamental algorithm, which is named the implicit Gauss–Cholesky algorithm, since it provides (implicitly) the same LU or LL^T factorization of A that is provided by these classical algorithms. One of the interesting properties of this algorithm is that it generates, with a low overhead, A-conjugate directions even without the requirement that A is positive definite. This property may lead to applications in the field of nonlinear optimization.

In Chapter 7 we apply the basic ABS algorithm to the scaled system $V^T A x = V^T b$, obtaining an essential generalization of the algorithms, called the scaled ABS algorithms, where the column vectors v_i of the nonsingular matrix V play the role of extra free parameters. We extend to the scaled ABS class the general properties and alternative formulations considered in Chapters 3 and 4 and we give additional results. We also provide a block formulation of the class that will allow a very general formulation of the ABS algorithms for nonlinear systems. Finally we investigate the relationships with the class of algorithms proposed by Stewart (1973) and Broyden (1985).

In Chapter 8 we consider subclasses of the scaled ABS algorithm corresponding to various choices of the scaling matrix V and we study particular algorithms obtained by special selection of the other free parameters. We derive the ABS formulation of several known algorithms, among them the QR, Hestenes–Stiefel, the Lanczos, the Craig and the Fridman methods. It is also shown that the ABS algorithm has potential applications in the eigenvalue–eigenvector field, which are not developed in this monograph.

In Chapter 9 we consider how to determine a minimal Euclidean norm least squares solution of overdetermined linear systems. We describe various approaches where the Huang algorithm plays a fundamental role.

Chapter 10 presents numerical experiments of several versions of the Huang algorithm on ill-conditioned linear systems and linear least squares (well conditioned, ill-conditioned and rank-deficient). The so-called modified Huang algorithm is shown to be an efficient algorithm, able to provide in most cases a more accurate estimate of the solution than that given by some commercial codes.

In Chapter 11 we consider the application of some ABS algorithms to special linear systems, which are expected in practice to be of large size. We show that the ABS algorithms adapt naturally to the structure of these problems and in some cases require less storage or overhead than the standard formulation of classical algorithms.

In Chapter 12 we provide a sensitivity analysis of the ABS class, following the Broyden (1974) model, and a standard forward error analysis for a certain computation model. From the sensitivity analysis we get a further indication in favour of the Huang algorithm, which has the following property: under the conditions of the Broyden model, errors in the estimate of the solution and in the residual cannot be amplified proportionally to the condition number of A. The forward error analysis is not of great practical use since it gives very general and pessimistic bounds.

Chapter 13 formulates the scaled block ABS algorithm for nonlinear systems. This is a very general class which contains as special cases the Newton, the Brown and the Brent methods. We give a local convergence result under weak conditions on the free parameters, which is an extension of the classical result for the Newton method. Limited numerical experiments indicate that the nonlinear modified Huang algorithm performs virtually identically with the very efficient implementation of the Brent method developed by Moré and Cosnard (1979).

1.3 NOTATION AND SPECIAL DEFINITIONS

We list the notation used in the monograph and we introduce some special definitions which are not standard in the literature.

Real numbers, also called 'scalars', are indicated in general by lower-case Greek letters. The exception to this rule is integers which are used as indexes, indicated by lower-case Roman letters, and the components of vectors and matrices, which are indicated by attaching indexes to the vector and matrix symbols.

Vectors are indicated by lower-case Roman letters; their components are real numbers. The set of all vectors of n components is indicated by R^n.

Matrices are indicated by capital Roman letters and their components are real numbers. The set of all matrices having n rows and m columns is indicated by $R^{n,m}$. The element of the matrix $A \in R^{m,n}$ belonging to the ith row and jth column is indicated by $A_{i,j}$. A vector is also considered as a one-column matrix in $R^{n,1}$, implying that $R^1 = R^{n,1}$.

The angle between two vectors a, b is indicated by $(a,b) \sphericalangle$.

By the symbol ^{T} we indicate transposition. If a is a column vector in $R^{n,1}$, by a^T we indicate the row vector in $R^{1,n}$.

By I_n we indicate the unit matrix in $R^{n,n}$; if there is no danger of confusion, the index n may be omitted. By $e_i \in R^n$ we indicate the ith unit vector in R^n. The ith component of a vector a can be indicated by $a^T e_i$ or by $a_{(i)}$.

We indicate sets by capital Greek letters, with the exception of the sets R^n, $R^{m,n}$.

The symbol $\| . \|$ indicates the norm of the enclosed arguments. Unless otherwise stated, the norm used for vectors is the Euclidean norm, $\|a\| = (a^T a)^{1/2}$. For matrices we consider the Frobenius, the weighted Frobenius and the vector-induced norm, specifying each time which norm is considered.

If a matrix A is square and nonsingular, then we indicate by A^{-1} its inverse. We indicate by A^{-T} the inverse of A^T, if it exists. We indicate by $A^{i,i}$ the ith principal submatrix of A, i.e. the matrix comprising the elements belonging to the intersection of the first i rows and columns. A square matrix A is called by us strongly nonsingular if all its principal submatrices are nonsingular. By A^+ we indicate the Moore–Penrose pseudoinverse of $A \in R^{n,m}$, i.e. the unique matrix A^+ satisfying the four relations $AA^+A = A$, $A^+AA^+ = A^+$, $AA^+ = (AA^+)^T$, $(A^+A) = (A^+A)^T$.

If A is square nonsingular, the condition number is defined by $\text{Cond}(A) = \|A\| \|A^{-1}\|$, once the particular matrix norm is specified. For A square singular or rectangular we consider the definition $\text{Cond}(A) = \|A\| \|A^+\|$.

The determinant, the rank, the trace, the range space and the null space of A are indicated by $\det(A)$, $\text{rank}(A)$, $\text{Tr}(A)$, $\text{Range}(A)$, and $\text{Null}(A)$.

Partitions of A are indicated by specific definitions of upper indexed matrices or by Householder (1964) notation, which is as follows: given $A \in R^{m,n}$ and $i < n$, $A^{|i}$ and $A^{i|}$ indicate respectively the matrix containing the first and the last i columns of A; given $k < m$, A^k indicates the matrix containing the first k rows of A.

The symbols $\in$, $\subset$, $\cup$ and $\cap$ indicate as usual inclusion, set inclusion, set union and set intersection. The symbol $\perp$ indicates the orthogonal complement to a space, as in the identity $\text{Range}(A) = {}^\perp\text{Null}(A^T)$.

By $\text{Span}(a_1, \ldots, a_i)$ we indicate the set of vectors which are linear combination of $a_1, \ldots, a_i$. By $b + \text{Span}(a_1, \ldots, a_i)$ we indicate the linear variety whose elements are obtained by adding to b a linear combination of $a_1, \ldots, a_i$.

By $K(a,A)$ we indicate the Krylov space associated with the vector a and the matrix A.

By $\mu(A)$ we indicate the spectral radius of the matrix A.

If $F(x)$ is a mapping from $x \in R^n$ to R^1, we indicate by $\delta F/\delta x_j$ the partial derivative of F with respect to the jth component of x and by $\text{grad}[F(x)]$ the gradient vector of F, i.e. the vector whose components are the partial derivatives of F.

2

Derivation of the ABS algorithm (basic form)

2.1 INTRODUCTION

The ABS algorithm in its basic form computes a solution x^+ of the system $Ax = b$ of m linear equations in n unknowns $(m \leq n)$ as the $(m + 1)$th iterate of a sequence of approximations x_i to x^+ having the following property: the approximation x_{i+1} obtained at the ith iteration is a solution of the first i equations.

The idea of solving linear systems by generating approximations with the above property is not new. Perhaps the oldest method based upon this idea is the so-called escalator method (see for instance Morris (1946) or Householder (1964)). A brief presentation of such a method is as follows. By obvious partitioning of A and x the subsystem consisting of the first i equations has the form

$$A^{i,i}x^i + A^{i,n-i}x^{n-i} = b^i . \tag{2.1}$$

If it is assumed that all the principal submatrices of A are nonsingular (a property that is always satisfied after suitable interchanges of the rows or the columns of A) the solution $(x^i)^+$ of the first i equations can be written formally as

$$(x^i)^+ = (A^{i,i})^{-1}(b^i - A^{i,n-i}x^{n-i}) , \tag{2.2}$$

where arbitrary values are assigned to x^{n-i}. In the escalator method, x^{n-i} is set to zero (thereby generating a sequence of basic-type solutions) and $(A^{i,i})^{-1}$ is computed from the knowledge of $(A^{i-1,i-1})^{-1}$ either by updating triangular factors or by using well-known formulas for the inverse of bordered matrices. In practice the escalator method is not recommended, because it has higher computational cost than other methods and the formulas for the updating of the inverse are prone to instability.

It will be shown later that the iterates x^i generated by the escalator method are identical (in exact arithmetic) with those obtained by the so-called implicit LU algorithm of the ABS class, when this is started with the zero vector as initial estimate for x^+.

Among other methods based upon the idea of solving at the ith step the first i equations are those of Pyle (1964), Huang (1975) and Sloboda (1978) and, in the context of solving nonlinear equations, those of Brown (1969), Brent (1973), Cosnard (1975) and Gay (1975a). They will be discussed later and their connections with the ABS class will be clarified.

2.2 HEURISTIC DERIVATION OF THE ABS ALGORITHM

We now present the heuristics behind the ABS algorithm, following Abaffy (1979). The idea is to look for an algorithm whose structure is basically that of the successful quasi-Newton algorithms for nonlinear systems of equations or for nonlinear unconstrained optimization (see for a review of such algorithms Spedicato (1976), Dennis and Moré (1977) or Dennis and Schnabel (1983)). A quasi-Newton-like algorithm has the following structure:

(A) Initialization phase: assign an arbitrary estimate of the solution, $x_1 \in R^n$; assign an arbitrary nonsingular matrix $H_1 \in R^{n,n}$; set the iteration index i equal to one.

(B) Determination of a search vector: compute a vector $p_i \in R^n$ by

$$p_i = H^T_i z_i \, , \tag{2.3}$$

where $z_i \in R^n$ is a vector whose specification will be discussed later.

(C) Update of the estimate of the solution: take a step along p_i from the current estimate x_i, say

$$x_{i+1} = x_i - \alpha_i p_i \, , \tag{2.4}$$

where α_i is a scalar, called the step size, whose value will be specified later.

(D) Stop if some convergence test for the most recent iterate x_{i+1} is satisfied; otherwise proceed.

(E) Update the matrix H_i by

$$H_{i+1} = H_i + D_i \, , \tag{2.5}$$

where the specific form for D_i will be given later.

(F) Increment the index i by one and go to (B).

We show in the following that the values of z_i, α_i and D_i can be determined by the requirement that the iterate x_{i+1} be a solution of the first i equations. More precisely, we demonstrate the following propositions.

(a) The vector z_i is essentially arbitrary but not orthogonal to a certain vector.

(b) The scalar α_i is determined by the ith equation.

(c) The correction D_i can be taken as a rank-one matrix containing a vector $w_i \in R^n$, which is essentially arbitrary, apart from the condition that its scalar product with a certain vector be equal to one.

The resulting algorithm contains therefore three essentially arbitrary quantities, say the initial matrix H_1 and the parameters z_i and w_i, $i = 1, \ldots, p$, where p is the index at which the algorithm is terminated. Thus it can be considered as a class of algorithms, since the parameters must be specified for an actual implementation. Particular choices will be studied later, showing *inter alia* that new formulations are obtained of classical methods such as the Gauss, the Cholesky, and the Gram–Schmidt methods.

In order to validate propositions (a)–(c), we proceed by induction. Let us assume that x_i has been determined so that the first $i - 1$ equations are satisfied. We have to determine z_i, α_i and D_i so that x_{i+1} solves the first i equations or, equivalently, so that the first i components of the residual $r(x_{i+1})$ are zero. First we observe that (2.4) implies for the residual $r(x)$ the identity

$$R_{(i)}(x_{i+1}) = r_{(i)}(x_i) - \alpha_i a_i^T p_i \ . \tag{2.6}$$

From (2.6) it follows that the ith equation is satisfied at x_{i+1} if $a_i^T p_i \neq 0$ and the step size is defined by

$$\alpha_i = \frac{a_i^T x_i - b_i}{a_i^T p_i} \ . \tag{2.7}$$

Since $p_i = H_i^T z_i$, the condition $a_i^T p_i \neq 0$ in terms of the vector z_i has the form

$$a_i^T H_i^T z_i \neq 0 \ . \tag{2.8}$$

Condition (2.8) can always be satisfied if $H_i a_i \neq 0$ by taking z_i arbitrary but nonorthogonal to $H_i a_i$. Since H_1 is nonsingular, $H_1 a_1$ is nonzero. For $i > 1$ the proof that $H_i a_i$ is nonzero depends on the structure of the update H_i, which is undefined at this stage. It will be shown in the next section that, with the update formula (2.15), the vector $H_i a_i$ is nonzero if and only if the ith row of A is not a linear combination of the first $i - 1$ rows.

We consider now the requirement that also the first $i - 1$ equations be zero. From the identity

$$r_{(j)}(x_{i+1}) = r_{(j)}(x_i) - \alpha_i a_j^T p_i \ , \qquad j \leqslant i - 1 \ , \tag{2.9}$$

and the induction assumption, it follows that $r_{(j)}(x_{i+1}) = 0$ if and only if

$$\alpha_i a_j^T H_i^T z_i = 0 \ . \tag{2.10}$$

The above condition is satisfied, disregarding the particular case when $\alpha_i = 0$, either when z_i is orthogonal to $H_i a_j$, $j = 1,\ldots,i-1$ or when H_i satisfies $H_i a_j = 0$, $j = 1,\ldots,i-1$. We consider the second approach. Assuming by induction that $H_i a_j = 0$ for $j \leqslant i-1$, (2.10) is satisfied and we have to determine the correction D_i so that $H_{i+1} a_j = 0$ for $j \leqslant i$. Proceeding in the spirit of the quasi-Newton updates, we look for a correction D_i of the lowest possible rank. Taking D_i as a rank-one matrix, we get

$$H_{i+1} = H_i + c_i d_i^{\mathrm{T}} . \tag{2.11}$$

For $j = i$ we obtain

$$H_i a_i + c_i d_i^{\mathrm{T}} a_i = 0 , \tag{2.12}$$

which is satisfied when $H_i a_i$ is nonzero if and only if $c_i = -\delta H_i a_i$ and $d_i^{\mathrm{T}} a_i = 1/\delta$, with δ an arbitrary nonzero scalar. Taking $\delta = 1$ the update becomes

$$H_{i+1} = H_i - H_i a_i d_i^{\mathrm{T}} . \tag{2.13}$$

For $j < i$ the condition $H_{i+1} a_j = 0$ implies, from the assumptions $H_i a_j = 0$ and $H_i a_i \neq 0$, the equations $d_i^{\mathrm{T}} a_j = 0$. These equations are satisfied if we set $d_i = H_i^{\mathrm{T}} w_i$, with $w_i \in R^n$ arbitrary, save for the condition

$$w_i^{\mathrm{T}} H_i a_i = 1 , \tag{2.14}$$

which follows from the relation $d_i^{\mathrm{T}} a_i = 1$. Condition (2.14) can always be satisfied by taking w_i as a suitable multiple of any vector nonorthogonal to $H_i a_i$. The final update formula for H_i is therefore

$$H_{i+1} = H_i - H_i a_i w_i^{\mathrm{T}} H_i , \tag{2.15}$$

with w_i arbitrary subject to (2.14). Finally, it is straightforward to start the induction by showing that $r_{(1)}(x_2) = 0$ and $H_2 a_1 = 0$ whenever x_i and H_i are updated according to (2.4), (2.7) and (2.15), with z_1, w_1 satisfying (2.8) and (2.14).

A geometrical interpretation of the algorithm is the following. At the ith step the vector x_i belongs to the $(n-i+1)$th dimensional linear variety Φ^{n-i+1} which contains all solutions of the first $i-1$ equations. The step along p_i defined by (2.4) leads to a new point x_{i+1} in the linear variety Φ^{n-i} which is included in Φ^{n-i+1}. Since a_i is a vector orthogonal to the hyperplane Σ^i, containing all solutions of the ith equations, condition (2.8) simply means that p_i must not be parallel to Σ^i, so that the intersection of Σ^i with the straight line $x(\alpha) = x_i - \alpha p_i$ occurs at the unique finite point $x_{i+1} = x(\alpha_i)$. Since p_i is essentially arbitrary in the range of H_i^{T}, the range of H_i^{T} must be included in $(\overline{\Phi})^{n-i+1}$, the translation of Φ^{n-i+1} which includes the origin. Actually it will be shown in the next chapter that Range(H_i^{T}) coincides with $(\overline{\Phi})^{n-i+1}$.

2.3 BIBLIOGRAPHICAL REMARKS

Iterations of the form (2.4) with the step size (2.7) have appeared in the literature also in the context of iterative algorithms for linear systems (see for instance Kaczmarz (1937), Cimmino (1938), Tanabe (1971) and Sloboda (1988)). The update formula (2.15) with $H_1 = I$, $w_i = a_i/a_i^T H_i a_i$ is the well-known orthogonal projection formula, which has been applied in various contexts. The first algorithm belonging formally to the ABS class appears to be the algorithm of Huang (1975), which corresponds to the choices $H_1 = I$, $z_i = a_i$, $w_i = a_i/a_i^T H_i a_i$. The ABS class was first given in a restricted form (with $H_1 = I$ and $z_i = a_i$) by Abaffy (1979). The introduction of the arbitrary matrix H_1 is due to Spedicato and has appeared in the papers by Abaffy and Spedicato (1982, 1984). The introduction of the parameter z_i is due to Broyden and has appeared in Abaffy *et al.* (1982, 1984b). The name ABS has been given to the class to acknowledge the contribution of Abaffy, Broyden and Spedicato in introducing the three parameters.

3

Basic properties of iterates of the ABS class

3.1 INTRODUCTION

In the previous chapter we have derived a class of algorithms, named the ABS class, for solving m linear equations in n unknowns ($m \leq n$). The class has been derived by analogy with the quasi-Newton methods. In exact arithmetic and under a certain condition (namely that the vectors $H_i a_i$ are nonzero) the algorithms are well defined and have the property that the iterate x_{m+1} solves the given equations. They are therefore algorithms of the direct type.

In this section we first reformulate the class in a slightly modified form, which is needed for dealing with the rank-deficient case. Then we prove a number of fundamental properties of the iterates H_i, p_i, x_i, completing, in particular, the validation of the algorithm.

The ABS class of algorithms is defined by the following procedure

ALGORITHM 1: The Basic ABS Algorithm
(A1) Let $x_1 \in R^n$ be arbitrary. Let $H_1 \in R^{n,n}$ be arbitrary nonsingular. Set $i = 1$ and $i$flag $= 0$.
(B1) Compute the ith component of the residual vector in x_i, say $\tau_i = a_i^T x_i - b^T e_i$ and the vector $s_i = H_i a_i$.
(C1) If $s_i \neq 0$ go to (D1); if $s_i = 0$ and $\tau i = 0$ set $x_{i+1} = x_i$, $H_{i+1} = H_i$, $i$flag $= i$flag $+ 1$ and go to (G1) if $i < m$; otherwise stop. If $s_i = 0$ and $\tau_i \neq 0$ set $i$flag $= -i$ and stop.
(D1) Compute the search vector p_i by

$$p_i = H_i^T z_i, \tag{3.1}$$

where $z_i \in R^n$ is arbitrary save for the condition

$$z_i^T H_i a_i \neq 0. \tag{3.2}$$

(E1) Update the approximation of the solution by

$$x_{i+1} = x_i - \alpha_i p_i,$$ (3.3)

where the step size α_i is given by

$$\alpha_i = \frac{\tau_i}{a_i^T p_i}$$ (3.4)

If $i = m$ stop; x_{m+1} solves the system.
(F1) Update the matrix H_i by

$$H_{i+1} = H_i - H_i a_i w_i^T H_i$$ (3.5)

where $w_i \in R^n$ is arbitrary save for the condition

$$w_i^T H_i a_i = 1.$$ (3.6)

(G1) Increment the index i by one and go to (B1).

The main difference with the algorithm introduced above is the step (C1) which checks whether $H_i a_i$ is zero or not. From Corollary 3.2 the vector $H_i a_i$ is zero if and only if a_i is a linear combination of $a_1, \ldots, a_{i-1}$. In such a case there are two possibilities: either the ith equation is a linear combination of the previous equations, or it is not, implying that the system is incompatible. Since x_i solves the first $i-1$ equation, discrimination between the two cases is obtained by checking whether τ_i is zero or not. In the first case it is natural to eliminate the ith equation, which results in setting $x_{i+1} = x_i$, $H_{i+1} = H_i$; in the second case it is natural to stop the algorithm. The exit parameter $i$flag gives information about the state of the system. If $i$flag $= 0$ no linear dependence of the rows of A has been detected. If $i$flag > 0, $i$flag equations are dependent. If $i$flag < 0, the system has no solution, since at least the equation of index $i = -i$flag is incompatible.

In the formulation of ALGORITHM 1 we have assumed that exact arithmetic is performed. In the presence of round-off the following questions are important.

— In step (C1), tolerances must be introduced in evaluating whether $H_i a_i$ and/or τ_i are zero or not.
— The best order to deal with the equations i.e. the optimal pivoting problem.

3.2 PROPERTIES OF THE MATRICES H_i

We consider now some fundamental properties of the iterates H_i, p_i, x_i generated by the ABS class. We start with the following theorem, which is basic for establishing that the class is well defined.

Theorem 3.1

Let $H_1 \in R^{n,n}$ be a nonsingular matrix and let $a_1, \ldots, a_m$ be linearly independent vectors in R^n ($m \leqslant n$). Consider, for $i = 1, 2, \ldots, m$, the sequence of matrices H_i generated by update (3.5), with w_i arbitrary vector in R^n. Then for $i \leqslant j \leqslant m$, the vectors $H_i a_j$ are nonzero and linearly independent.

Proof

We proceed by induction. For $i = 1$ the theorem is true since H_1 is nonsingular. Assuming that the theorem is true up to $i < m$, we prove it for $i + 1$. From (3.5) we have for $j \geqslant i + 1$

$$H_{i+1}a_j = H_i a_j - H_i a_i w_i^{\mathrm{T}} H_i a_j \tag{3.7}$$

which implies that $H_{i+1}a_j$ is nonzero, being the sum of two vectors, the first nonzero and the second either zero or linearly independent from the first. We now have to prove that relation

$$\sum_{k=i+1}^{m} \beta_k H_{i+1} a_k = 0 \tag{3.8}$$

implies that $\beta_k = 0$. Using (3.5) we can write (3.8) as follows:

$$\sum_{k=i+1}^{m} \beta_k H_i a_k - \left(\sum_{k=i+1}^{m} \beta_k w_i^{\mathrm{T}} H_i a_k \right) H_i a_i = 0 \tag{3.9}$$

or

$$\sum_{k=1}^{m} \delta_k H_i a_k = 0 , \tag{3.10}$$

with $\delta_k = \beta_k$ for $k \geqslant i + 1$, $\delta_i = - \sum_{k=i+1}^{m} \beta_k w_i^{\mathrm{T}} H_i a_k$. Since the induction implies that $\delta_k = 0$ for $k \geqslant i$, it follows that $\beta_k = 0$ for $k \geqslant i + 1$. Q.E.D.

Corollary 3.1

The vector $H_i a_i$ computed by the ABS algorithm is zero only if a_i is a linear combination of $a_1, \ldots, a_{i-1}$.

Theorem 3.2

Assume that rank$(A) = m$. Then the ABS algorithm is well defined.

Proof
We have to show that z_i and w_i can be chosen so that (3.2) and (3.6) are satisfied. Since by Theorem 3.1 the vectors $H_i a_i$ are nonzero, a vector u_i exists such that $u_i^T H_i a_i \neq 0$, for instance $u_i = H_i a_i$. Then the ABS algorithm is well defined if we take $z_i = \beta_i u_i$, with $\beta_i \neq 0$ and w_i according to

$$w_i = \frac{u_i}{u_i^T H_i a_i} \tag{3.11}$$

Q.E.D.

The proof of the following theorem is immediate by induction and is omitted (it has been essentially given in the previous chapter).

Theorem 3.3
Given m arbitrary vectors $a_1, \ldots, a_m \in R^n$ and an arbitrary nonsingular matrix $H_1 \in R^{n,n}$, consider the sequence of matrices $H_2, \ldots, H_{m+1}$ generated by (3.5) with w_i satisfying (3.6) if such a w_i exists, otherwise according to $H_{i+1} = H_i$. Then the following relations are true for $i = 2, \ldots, m+1$:

$$H_i a_j = 0 \qquad 1 \leq j < i. \tag{3.12}$$

Corollary 3.2
The vector $H_i a_i$ computed by the ABS algorithm is zero if and only if a_i is a linear combination of $a_1, \ldots, a_{i-1}$.

Corollary 3.3
The ABS algorithm is well defined for arbitrary rank of A.

Unless otherwise stated, we shall tacitly assume in the sequel that A is full rank, say rank$(A) = m$. This assumption is made essentially to simplify the formulation of the studied properties, whose extension to the general case is usually trivial.

Theorem 3.4
For $1 \leq i \leq m+1$ the rank of the matrices H_i, defined by update (3.5) with H_1 nonsingular and w_i satisfying (3.6), is equal to $n - i + 1$.

Proof
Consider first the case where $m = n$. Then Theorem 3.1 implies that dim[Range(H_i)] $\geq n - i + 1$, while Theorem 3.3 implies that dim[Null(H_i)] $\geq i - 1$. Since dim[Range(H_i)] + dim[Null(H_i)] $= n$ it follows that the equality sign holds in the previous inequalities, and the theorem is true. If $m < n$ add $n - m$ linearly independent vectors to $a_1, \ldots, a_m$ and apply the previous result. Q.E.D.

Corollary 3.4

Consider the matrices H_i defined by update (3.5) with H_1 nonsingular and w_i satisfying (3.6). Then, for $i = 2, \ldots, m+1$, Null(H_i) is spanned by $a_1, \ldots, a_{i-1}$. If $m = n$, then Range(H_i) is spanned by $H_i a_i, \ldots, H_i a_n$.

The proof of the following theorem is omitted, being an immediate consequence of the previously given results.

Theorem 3.5

Consider the matrices H_i generated by the ABS algorithm; let q be the rank of A, $1 \leqslant q \leqslant m$; assume that the system $Ax = b$ is compatible; let q_i be the number of times that the equality $H_i a_j = 0$ occurs for $j \leqslant i$. Then the following statements are true for $i = 1, \ldots, m$.

(a) rank(H_{i+1}) = rank(H_i) $- 1$ if $i = 1$ or $i > 1$ and a_i is not a linear combination of $a_1, \ldots, a_{i-1}$.

(b) rank(H_{i+1}) = rank(H_i) if $i > 1$ and a_i is a linear combination of $a_1, \ldots, a_{i-1}$.

Moreover the following identities are true:

$$\text{rank}(H_{i+1}) = n - i + q_i, \tag{3.13}$$

$$q = m - q_m. \tag{3.14}$$

If moreover $m = n$ and $q = m$,

$$H_{n+1} = 0. \tag{3.15}$$

The following theorem establishes projection type properties of the matrices H_i.

Theorem 3.6

Consider the matrices H_i generated by (3.5) with H_1 nonsingular and w_i satisfying (3.6). Then for $i = 1, \ldots, m+1$ and $1 \leqslant j \leqslant i$, the following relations are true:

$$H_i H_1^{-1} H_j = H_i, \tag{3.16}$$

$$H_j H_1^{-1} H_i = H_i. \tag{3.17}$$

Proof

We only prove (3.16) since the proof for (3.17) is similar. We proceed by induction. For $i = 1$, (3.16) is trivially true. Assuming now that (3.16) is true up to the index i, we have to prove that $H_{i+1} H_1^{-1} H_j = H_{i+1}$. For $j = i + 1$, we have $H_{i+1} H_1^{-1} H_{i+1} = (H_i - H_i a_i w_i^T H_i) H_1^{-1} (H_i - H_i a_i w_i^T H_i) = H_i - 2 H_i a_i w_i^T H_i + H_i a_i w_i^T H_i w_i^T H_i a_i = H_{i+1}$ by (3.6) and the induction. For $j < i + 1$ we have $H_{i+1} H_1^{-1} H_j = (H_i - H_i a_i w_i^T H_i) H_1^{-1} H_j = H_i - H_i a_i w_i^T H_i = H_{i+1}$ again by (3.6) and the induction. Q.E.D.

Remark 3.1
If $H_1 = I$, Theorem 3.6 implies that the matrices H_i are idempotent or projection matrices. For arbitrary H_i it is easily seen that, if we define $\overline{H}_i$ by

$$\overline{H}_i = H_1^{-1}H_i,\tag{3.18}$$

then the matrices $\overline{H}_i$ are idempotent or projection matrices.

Theorem 3.7
Let H_1 be symmetric positive definite and consider the matrices H_i generated by (3.5) with w_i some vector. Then the following statements are true for $i = 1, \ldots, m$.

(a) The following formula for w_i is well defined and satisfies (3.6):

$$w_i = \frac{a_i}{a_i^T H_i a_i}.\tag{3.19}$$

(b) The denominator in (3.19) is strictly positive.
(c) The matrix H_{i+1} is symmetric.

Proof
We proceed by induction. For $i = 1$, $a_1^T H_1 a_1$ is positive since H_1 is symmetric positive definite, so that (3.19) is well defined. It is also immediate to verify that (3.6) is satisfied and that H_2 is symmetric. Assuming now the validity of the statements up to the index $i - 1$, we have that $a_i^T H_i a_i = a_i^T H_i^T H_1^{-1} H_i a_i = \|H_1^{-1/2} H_i a_i\| > 0$, by use of symmetry and of the Theorems 3.6 and 3.1. Therefore (b) is true and (3.19) is well defined. It is also immediate to verify that (3.6) is satisfied and that H_{i+1} is symmetric. Q.E.D.

Remark 3.2
Equation (3.19) for w_i was used by Huang (1975), with $H_1 = I$ and $z_i = a_i$, in the paper which led to the development of the ABS class.

Remark 3.3
Equation (3.19) for w_i is not the only one which generates a sequence of symmetric matrices H_i. For instance it follows from Theorem 3.6 that, if $H_1 = I$ and S_k has the form

$$S_k = \prod_{j=1}^{k} T_j,\tag{3.20}$$

where T_j is either H_i or H_i^{T} and k is arbitrary, then

$$w_i = \frac{S_k a_i}{a_i^{\mathrm{T}} S_k H_i a_i} \tag{3.21}$$

satisfies (3.6) and generates the same update as (3.19).

Definition 3.1
The update where H_1 is symmetric positive definite and w_i is chosen according to (3.19) is called the symmetric update. The update where $H_1 = I$ and w_i is chosen according to (3.19) is called the Huang update.

 We now consider some properties involving the transpose of H_i. Observing that H_i^{T} is updated by

$$H_{i+1}^{\mathrm{T}} = H_i^{\mathrm{T}} - H_i^{\mathrm{T}} w_i a_i^{\mathrm{T}} H_i^{\mathrm{T}}, \tag{3.22}$$

it is clear that any property of H_i of the form $Q(H_i, a_i, w_i)$ obtained under the condition that the sequence a_i has property P1 and the sequence w_i has property P2 can be restated as a property of H_i^{T} of the form $Q(H_i^{\mathrm{T}}, w_i, a_i)$ under the condition that the sequence w_i has property P1 and the sequence a_i has property P2.

Theorem 3.8
Consider the matrices H_i generated by (3.5) with w_i satisfying (3.6) and H_1 nonsingular. Then the following statements are true for $i = 1, \ldots, m$.

(a) The vectors $w_1, \ldots, w_i$ are linearly independent.
(b) The vectors $w_1, \ldots, w_i$ span $\mathrm{Null}(H_{i+1}^{\mathrm{T}})$.

Proof
We proceed by induction. For $i = 1$, w_1 is nonzero since H_1^{T} is nonzero and (3.6) is satisfied; it is moreover immediate to verify that $H_2^{\mathrm{T}} w_1 = 0$. Assuming now the validity of the statements up to $i - 1$, we observe that, if w_i were a linear combination of $w_1, \ldots, w_{i-1}$, then $H_i^{\mathrm{T}} w_i = 0$ by induction, implying that $w_i^{\mathrm{T}} H_i a_i = 0$, against the assumption; thus $w_1, \ldots, w_i$ are linearly independent. It is immediate to verify that $H_{i+1}^{\mathrm{T}} w_i = 0$ from (3.6) and that $H_{i+1}^{\mathrm{T}} w_j = 0$ for $1 \leqslant j < i$ by induction. Q.E.D.

 The proof of the following theorem is omitted, since it is the restatement in terms of the transpose of Theorem 3.1.

Theorem 3.9
Let $H_1 \in R^{n,n}$ be a nonsingular matrix and let $w_1, \ldots, w_m$ be linearly independent vectors in R^n, $m \leqslant n$. Consider, for $i = 1, \ldots, m$, the sequence of matrices H_i generated by (3.5), with $a_i \in R^n$ arbitrary. Then, for $i \leqslant j \leqslant m$, the vectors $H_i^{\mathrm{T}} w_j$ are nonzero and linearly independent.

Remark 3.4
Define the following matrices:

$$A^i = (a_1, \ldots, a_i),$$
(3.23)

$$A^{m-i} = (a_{i+1}, \ldots, a_m),$$
(3.24)

$$W^i = (w_1, \ldots, w_i),$$
(3.25)

$$W^{m-i} = (w_{i+1}, \ldots, w_m).$$
(3.26)

Then the previously proved results can be expressed by the following relations:

$$H_{i+1}A^i = 0,$$
(3.27)

$$H_{i+1}{}^T W^i = 0.$$
(3.28)

Moreover, if $\text{rank}(A^i) = i$,

$$\text{rank}(W^i) = i,$$
(3.29)

$$\text{rank}(H_{i+1}A^{m-i}) = m - i,$$
(3.30)

$$\text{rank}(H_{i+1}{}^T W^{m-i}) = m - i.$$
(3.31)

Remark 3.5
If $\text{rank}(A^i) = i$, $m = n$ and $H_1 = I$, then the columns of A^i are eigenvectors of H_{i+1} corresponding to the eigenvalue zero with multiplicity i while the columns of $H_{i+1}A^{n-i}$ are eigenvectors of H_{i+1} corresponding to the eigenvalue 1 with multiplicity $n - i$. The eigenvectors corresponding to eigenvalue zero are orthogonal to those corresponding to the eigenvalue one. The corresponding result for the eigenvectors of $H_{i+1}{}^T$ is similarly obtained in terms of W^i and $H_{i+1}{}^T W^{n-i}$. Finally it can be proved, see Gu (1988), that if the k_1th, . . . ,k_ith rows of W^i are linearly independent, then the $n - i$ rows of H_{i+1} not including the k_1th, . . . ,k_ith rows are also independent.

Theorem 3.10
Consider the matrices H_i defined by (3.5) with H_1 nonsingular and w_i satisfying (3.6). Consider the matrices $\overline{A}^i$, $\overline{W}^i$ defined by relations

$$\overline{A}^i = (H_1 a_1, \ldots, H_i a_i)$$
(3.32)

$$\overline{W}^i = (H_1{}^T w_1, \ldots, H_i{}^T w_i).$$
(3.33)

Then the following statements are true.

(a) $\overline{A}^i$ is full rank.
(b) $\overline{W}^i$ is full rank.
(c) If H_1 is symmetric then the columns of $\overline{A}^i$ and $\overline{W}^i$ are mutually H_1^{-1}-conjugate.
(d) If $H_1 = I$ and $m = n$, then $\overline{A}^n$ is the inverse of $\overline{W}^n$.

Proof
From Theorem 3.3 and equation (3.6) it follows that the scalar product $w_i^T H_i a_j$ equals zero for $j < i$ and one for $j = i$; in matrix terms this implies $\overline{W}_i^T A^i = R_1$ where R_1 is an upper unit triangular matrix. Similarly from Theorem 3.8 and equation (3.6) it follows that the scalar product $a_i^T H_i^T w_j$ equals zero for $j < i$ and one for $j = i$, implying $\overline{A}^{iT} W^i = R_2$ where R_2 is an upper unit triangular matrix. Since the ranks of R_1, R_2, A^i, W^i are equal to i, it follows that also $\overline{A}^i$ and $\overline{W}^i$ are full rank, proving (a) and (b). To prove (c) we observe that, if $i < j$, then $(H_i a_i)^T H_1^{-1}(H_j^T w_j) = a_i^T H_i^T w_j = 0$ from Theorem 3.3 and Theorem 3.6 while, if $i > j$ then $(H_i a_i)^T H_1^{-1}(H_j^T w_j) = a_i^T H_i^T w_j = 0$ from Theorem (3.8) and Theorem (3.6). Since $(H_i a_i)^T H_1^{-1}(H_i^T w_i) = a_i^T H_i^T w_i = 1$ from Theorem 3.6 and equation (3.6), we can write

$$\overline{A}^{iT} H_1^{-1} \overline{W}^i = I_i \tag{3.34}$$

which implies (d). Q.E.D.

The following theorem gives bounds to the Frobenius norm of H_i, which are independent of A and will be used in the convergence analysis of the nonlinear ABS methods (see Chapter 13).

Theorem 3.11

Let $H_1 = I$ and consider for $i = 1, \ldots, m$ the update (3.5) with $w_i = \dfrac{u_i}{u_i^T H_i a_i}$. Suppose that u_i is chosen so that one of the following conditions holds.

(a) The sequence H_i consists of symmetric matrices.
(b) The modulus of the angle between w_i and $H_i a_i$ is less than $\beta < \pi/2$.
(c) The modulus of the angle between $H_i^T w_i$ and a_i is less than $\alpha < \pi/2$.

Then the Frobenius norm of H_i is bounded by a constant depending only on n in case (a) and on m, n and β in cases (b) and (c).

Proof
Let $\|.\|$ indicate the Frobenius norm. In case (a), H_i is an orthogonal projector; hence $\|Hi\| \leq n^{1/2}$. In case (b) we have $\|H_{i+1}\| \leq \|H_i\|(\|I\| + \|H_i a_i w_i^T\|/|u_i^T H_i a_i|) - \|H_i\|(n^{1/2} + 1/\cos\beta)$; hence $\|H_{i+1}\| \leq n^{1/2}(n^{1/2} + 1/\cos\beta)^i \leq] - {}^\lq t - n^{1/2}(n^{1/2} + 1/\cos\beta)^m$. Case (c) is treated similarly to case (b) and results in the same bound. Q.E.D.

In quasi-Newton algorithms for nonlinear equations or nonlinear optimization, classes of updates for H_i have been constructed using a variational criterion whereby, in $H_{i+1} = H_i + D_i$, D_i is determined by minimizing a weighted Frobenius norm subject to conditions such as the quasi-Newton equation and symmetry (see for instance Greenstadt (1970), Spedicato and Greenstadt (1978) and Flachs (1982)).

This approach has been developed by Spedicato (1987a) in the context of the ABS methods in terms of the following weighted Frobenius norm of a matrix B:

$$\|B\|^2_{W,Q} = \text{Tr}(W^{-1}B^TQ^{-1}B), \tag{3.35}$$

where W and Q are symmetric positive definite matrices. Minimizing the norm (3.35) of the correction to H_i in (3.5) with respect to w_i under the condition (3.6) and with $H_1 = I$ leads then to the following update, which depends only on W as a parameter and where $e_i \in R^i$:

$$H_{i+1} = H_i - H_i a_i e_i^T[(A^i)^TWA^i]^{-1}(A^i)^TW. \tag{3.36}$$

It can be shown (see also Abaffy and Spedicato (1983b)) that the Huang update corresponds to setting $W = I$ in (3.36). It can also be shown, (Deng and Xiao 1989) that every ABS update of H_i satisfies a variational criterion with respect to a suitable W.

3.3 PROPERTIES OF THE SEARCH VECTORS

We consider now some properties of the search vectors p_i. Since we are assuming, unless otherwise stated, that $\text{rank}(A) = m$, each iteration defines a new search vector.

The following theorem plays a fundamental role in the analysis of the ABS class.

Theorem 3.12
Consider for $i \leqslant m$ the matrix A^i defined by (3.23) and define the matrix P^i by

$$P^i = (p_1, \ldots, p_i). \tag{3.37}$$

Then the matrix L^i defined by

$$L^i = (A^i)^TP^i \tag{3.38}$$

is lower triangular and nonsingular.

Proof
As $(L^i)_{j,k} = a_j^Tp_k = (H_k a_j)^Tz_k$, Theorem 3.3 implies that $(L^i)_{j,k}$ is zero for $j < k$, so that L^i is lower triangular. Since $a_i^Tp_i = a_i^TH_i^Tz_i$ is nonzero from (3.2), L^i is nonsingular. Q.E.D.

Corollary 3.5
The vectors $p_1, \ldots, p_i$ are nonzero and linearly independent.

Corollary 3.6
The vector p_i is parallel to the hyperplanes $\Sigma^1, \ldots, \Sigma^{i-1}$.

Corollary 3.7
If $m = n$ the matrix A can be factorized in the form

$$A = LS \tag{3.39}$$

with $L = L^n$ and $S = (P^n)^{-1}$.

Remark 3.6
The factorization (3.39) is performed implicitly by the algorithms of the ABS class, since the matrix L is not actually formed (apart from its diagonal elements) nor is the matrix P stored or inverted. If the matrix L is formed and P is stored, then the solution of $Ax = b$, can be found by first solving the triangular system

$$Ly = b \tag{3.40}$$

and then computing x^+ by

$$x^+ = Py. \tag{3.41}$$

The following theorem shows that the parameters H_1 and w_i are redundant, in the sense that any sequence of search vectors p_i, and therefore of approximations x_i to the solution, can be obtained by using only the parameter z_i.

Theorem 3.13
The set of vectors p_i that can be defined at step (D1) of the ABS algorithm is independent of H_1 and of W^{i-1}.

Proof
The set of vectors p_i defined by equation (3.1) for all possible vectors z_i coincides with the range of H_i^T, which is the orthogonal complement of $\text{Null}(H_i)$. Since $\text{Null}(H_i)$ is spanned by $a_1, \ldots, a_{i-1}$ then $\text{Range}(H_i^T)$ depends only on A^{i-1} and not on H_1 and W^{i-1}, from which H_i depends. Condition (3.2) only eliminates those vectors in $\text{Range}(H_i^T)$ which are orthogonal to a_i, hence the set of vectors p_i defined at step (D1) depends only on A^i. Q.E.D.

Corollary 3.8
The set Φ^i of all possible vectors p_i defined at step (D1) of ALGORITHM 1 satisfies the relation

$$\text{Range}(A^{i-1}) = \Phi^i \cup [\text{Null}(a_i^T) \cap \text{Range}(A^{i-1})]. \tag{3.42}$$

Remark 3.7
Theorem 3.13 implies that no loss of generality is suffered if H_1 and w_i are taken according to a fixed choice (for instance $H_1 = I$ and $w_i = a_i/a_i^T H_i a_i$). However,

availability of degrees of freedom in H_1 and w_i is important from the following points of view.

— The numerical stability of the update formula (3.5) can depend on H_1 and w_i.
— The computational costs of the update formula (3.5) can depend on H_1 and w_i.

3.4 PROPERTIES OF THE APPROXIMATION x_i TO THE SOLUTION

We conclude this section by considering some properties of the iterates x_i. First we state in the general form the basic property of the ABS algorithms that has been used in Chapter 2 to derive ALGORITHM 1.

Theorem 3.14

Let A be an m by n matrix, with $m \leq n$ and nonzero rows; consider the application of any algorithm of the ABS class to the solution of the system $Ax = b$; suppose that the index $i$flag is nonnegative up to the iteration index $\bar{\imath}$, $1 \leq \bar{\imath} \leq m$. Then for all i such that $1 \leq i \leq \bar{\imath}$, the vector x_{i+1} solves the first i equations, say

$$a_j^T x_{i+1} = b^T e_j, \qquad 1 \leq j \leq i. \tag{3.43}$$

Proof

Since the index $i$flag is negative if and only if at least one equation is incompatible, the first $\bar{\imath}$ equations are compatible by assumption. We proceed by induction. For $i = 1$, $H_1 a_1$ is nonzero and (3.43) is satisfied by the choice of α_1, as shown in Chapter 2. Assume now the validity of the theorem up to $k - 1$, $k \leq \bar{\imath}$, so that x_k solves the first $k - 1$ equations. If $H_k a_k$ is nonzero, then (3.43) is satisfied for $j = k$ by the choice of α_k, for $j < k$ by the induction and the relation $H_k a_j = 0$, (see Chapter 2). If $H_k a_k = 0$, since then $x_{i+1} = x_i$ the first $k - 1$ equations are trivially satisfied by induction, while the kth equation is satisfied since i flag is nonnegative. Q.E.D.

The following theorem gives an explicit representation of the linear variety which contains all the solutions of the first i equations.

Theorem 3.15

Assume that the conditions of the Theorem 3.13 are satisfied; let q_i be defined as in Theorem 3.5; let $\bar{x}$ be any vector of the form

$$\bar{x} = x_{i+1} + H_{i+1}^T s, \tag{3.44}$$

with s arbitrary in R^n. Then the following statements are true.
(a) $\bar{x}$ solves the first i equations of the given system, say

$$a_j^T \bar{x} = b^T e_j, \qquad 1 \leq j \leq i. \tag{3.45}$$

(b) The set Φ of all vectors $\bar{x}$ of the form (3.44) coincides with the linear variety containing all solutions of the first i equations.

Proof

Since a_j for $1 \leqslant j \leqslant i$ belongs to Null(H_{i+1}), (3.45) is true, since it reduces to (3.43). Let Ω be the linear variety, whose dimension is $n - i + q_i$, which contains all solutions of the first i equations. By definition, Φ is contained in Ω. Since $H_{i+1}{}^T$ has rank $n - i + q_i$, (see (3.14)), the dimension of Φ is $n - i + q_i$ and thus Φ coincides with Ω. Q.E.D.

Since the solution x^+ belongs to $\Omega = \Phi$, a vector s exists such that $\bar{x} = \bar{x}(s)$ in (3.43) coincides with x^+, say $\bar{x}$ solves not only the first i equations, but also the $m - i$ remaining ones. If we call such a vector s^+, then it is clear that it is not unique, as $H_{i+1}{}^T$ is singular; indeed all vectors s of the form

$$\bar{s} = s^+ + \sum_{j \in J} w_j \tag{3.46}$$

where J is the set of indexes $j \leqslant i$ such that $H_j a_j$ is nonzero, solve the system, since the vectors w_j belong to Null($H_{i+1}{}^T$).

The following theorem indicates how to compute a special vector s^+ such that $\bar{x}(s^+) = x^+$.

Theorem 3.16

Consider the system $Ax = b$ with $A \in R^{m,n}$ and full rank. Let $\bar{x} = \bar{x}(s)$ be defined as in (3.44) for $1 \leqslant i < m$. Then there exists a unique vector s^+ of minimal Euclidean norm such that the equation $A\bar{x}(s^+) = b$ is satisfied. Such a vector has the form

$$s^+ = Z^T d, \tag{3.47}$$

where Z is the full rank matrix defined by

$$Z = (H_{i+1} a_{i+1}, \ldots, H_{i+1} a_m)^T \tag{3.48}$$

and $d \in R^{m-i}$ satisfies the positive definite linear system

$$ZZ^T d = u, \tag{3.49}$$

with $u \in R^{m-i}$ the vector satisfying relations

$$u^T e_{j-i} = -(Ax_{i+1} - b)^T e_j, \qquad j = i+1, \ldots, m. \tag{3.50}$$

Proof

Since for any s the vector $\bar{x}(s)$ satisfies the first i equations, it is enough to require that the last $m - i$ equations be satisfied. They can be written in the form

$$a_j^{\mathrm{T}}H_{i+1}^{\mathrm{T}}s - (Ax_{i+1} - b)^{\mathrm{T}}e_j = 0 \qquad i+1 \leqslant j \leqslant m \tag{3.51}$$

or

$$Zs = u. \tag{3.52}$$

As Z is full rank, the underdetermined system (3.52) is compatible. Now the unique minimal Euclidean norm solution of an underdetermined linear system is a linear combination of the rows of the system, say

$$s^+ = \sum_{j=i+1}^{m} \delta_j H_{i+1} a_j. \tag{3.53}$$

Letting $d \in R^{m-i}$ be the vector defined by $d^{\mathrm{T}}e_{j-1} = \delta_j, j = i+1, \ldots, m$, (3.53) takes the form (3.47). Substituting (3.47) in (3.52), relation (3.49) is obtained, which defines d and then s^+ uniquely. Q.E.D.

Remark 3.8
If we write $s = -\alpha_{i+1}z_{i+1}$, (3.44) takes the form (3.2) written with the index $i+1$. Therefore Theorem 3.15 implies that, in step (D1) of ALGORITHM 1, z_i can always be chosen so that all the equations are satisfied in x_{i+1}.

The following theorem shows that at step (F1) of ALGORITHM 1 it is possible to choose w_i so that x_{i+2} solves all the equations, with z_{i+1} arbitrary save for a nonorthogonality condition and α_{i+1} equal to one.

Theorem 3.17
Consider the system $Ax = b$, with A m by n $(m \leqslant n)$ and full rank. Let $z \in R^n$ be any vector such that $a_i^{\mathrm{T}}H_i^{\mathrm{T}}z \neq 0$ (for instance $z = H_i a_i$). Then at the ith iteration it is possible to choose w_i such that at the $(i+1)$th iteration x_{i+2} solves the system with $z_{i+1} = z$ and $\alpha_{i+1} = 1$. Moreover the vector w_i^+ of minimal Euclidean norm having this property has the form

$$w_i^+ = Z^{\mathrm{T}}d, \tag{3.54}$$

where Z is full rank given by

$$Z = (H_i a_i, \ldots, H_i a_m)^{\mathrm{T}} \tag{3.55}$$

and $d \in R^{m-i+1}$ satisfies the positive definite linear system

$$ZZ^{\mathrm{T}}d = u, \tag{3.56}$$

with $u \in R^{m-i+1}$ the vector such that

$$u^T e_1 = 1 \tag{3.57}$$

and, for $j = 2, \ldots, m-i+1$,

$$u^T e_j = \frac{(-e_{i+j-1})^T r(x_{i+1} - H_i^T z)}{a_i^T H_i^T z}. \tag{3.58}$$

Proof
With the given definitions, x_{i+2} has the form

$$x_{i+2} = x_{i+1} - H_{i+1}^T z \tag{3.59}$$

and we note that, from Theorem 3.15, x_{i+2} solves the first i equations for any z. Substituting (3.5) in (3.59) the condition that the remaining $m-i$ equations be satisfied takes the form

$$e_j^T r(x_{i+1} - H_i^T z) + a_i^T H_i^T z a_j^T H_i^T w_i = 0, \qquad j = i+1, \ldots, m. \tag{3.60}$$

Since w_i must also satisfy (3.6), it is easily seen that (3.60) and (3.6) are equivalent to the following system:

$$Z w_i = u, \tag{3.61}$$

with $u \in R^{m-i+1}$ defined by (3.57) and (3.58). Since Z is full rank from Theorem 3.1, system (3.61) is compatible. By proceeding exactly as in the previous theorem it follows that the minimal Euclidean norm solution w_i^+ of (3.61) has the form given in the theorem. Q.E.D.

3.5 UPDATING THE SOLUTION OF A MODIFIED SYSTEM

One is often confronted with the following problem: given a solution x_{m+1} of the equations $a_j^T x = b^T e_j, j = 1, \ldots, m$, find a solution of the modified system, obtained by substituting the ith equation of the previous system by the equation $c^T x = d$. An answer to this problem is given in the following theorem, due to Zhu (1987). The proof is omitted for reasons of space.

Theorem 3.18
Suppose that x_{m+1} is a solution of the full rank system $a_j^T x = b^T e_j$, obtained by an ABS method generating the search vectors $p_1, \ldots, p_m$ and the matrix H_{m+1}, which are stored. Let J be the set of integers $1, \ldots, i-1, i+1, \ldots, m$. Consider the modified system defined by

$$a_j^T x = b^T e_j, \qquad j \in J, \tag{3.62}$$

$$c^T x = \sigma \tag{3.63}$$

Compute $\tau = c^T x_{m+1} - \sigma$. If $H_{m+1} c \neq 0$, compute $p_{m+1} = H_{m+1}^T z_{m+1}$, with $z_{m+1} \in R^n$ arbitrary save that $z_{m+1}^T H_{m+1} c \neq 0$, and $x_{m+2} = x_{m+1} - (\tau / c^T p_{m+1}) p_{m+1}$. Then the following statements are true.

(a) There exists a unique p of the form

$$p = p_i + \sum_{j=i+1}^{m+1} \beta_j p_j, \tag{3.64}$$

which satisfies equations $p^T c = 0$ and $p^T a_j = 0, j \in J$.

(b) The general solution of the modified system has the following form, with $s \in R^{n+1}$ arbitrary, $H_{m+2} = H_{m+1} - H_{m+1} c w_{m+1}^T H_{m+1}, w_{m+1}^T H_{m+1} \sigma = 1$

$$x = x_{m+2} + \begin{bmatrix} H_{m+2} \\ p^T \end{bmatrix}^T s. \tag{3.65}$$

Suppose that $H_{m+1} c = 0$. Then the following statements are true.

(c) There exists a unique vector p of the form

$$p = p_i + \sum_{j=i+1}^{m} \beta_j p_j \tag{3.66}$$

that satisfies equations $p^T a_j = 0, j = i+1, \ldots, m$.

(d) If $p^T c \neq 0$, the general solution of the modified system has the following form, with $s \in R^n$ arbitrary:

$$x = x_{m+1} - \frac{\tau}{p^T c} p + H_{m+1}^T s. \tag{3.67}$$

If $p^T c = 0$ and $\tau = 0$, the general solution of the modified system has the following form, with $s \in R^{n+1}$ arbitrary:

$$x = x_{m+1} + \begin{bmatrix} H_{m+1} \\ p^{\mathrm{T}} \end{bmatrix}^{\mathrm{T}} s. \qquad\qquad (3.68)$$

(b) If $p^{\mathrm{T}}c = 0$ and $\tau \# 0$, the modified system has no solution.

3.6 BIBLIOGRAPHICAL REMARKS

ALGORITHM 1 has appeared in the papers by Abaffy *et al.* (1982, 1984a,b). Theorems 3.1 and 3.2 have been given in Abaffy and Spedicato (1982), but the statement that $H_i a_i$ is nonzero has already been made by Huang (1975) for the particular case considered by him. Theorems 3.3 and 3.4 have appeared in the papers by Abaffy (1979) as generalization of results of Huang (1975). Theorems 3.5, 3.6, 3.12, 3.14 and 3.15 have been given by Abaffy and Spedicato (1982). Theorem 3.7 has appeared in the paper by Huang (1975). Theorems 3.8, 3.9 and 3.10 have appeared in the papers by Abaffy, *et al.* (1982, 1984a,b). Theorems 3.11 is from the papers by Abaffy, *et al.* (1987a,b). Theorems 3.13, 3.16 and 3.17 are new and due to Spedicato.

4

Alternative representations of the search vector and the matrix H_i

4.1 INTRODUCTION

In this chapter we consider alternative representations of the search vector and the matrix H_i, which may be preferable, in terms of storage requirement and/or computational cost, to the basic representation given in Chapter 3 (equations (3.1) and (3.5)). The first representation is mainly of theoretical interest, being useful in the analysis of properties of the ABS algorithm. The other representations are of interest in practical implementations of the algorithm. Some discussion of their relative merits and computational performance will be given later, with reference to special methods (mainly the Huang algorithm).

4.2 REPRESENTATION OF H_{i+1} IN TERMS OF H_1, A^i AND W^i

Theorem 4.1

Let $\overline{A}^i$ and $\overline{W}^i$ be defined as in Theorem 3.10. Then the matrix H_{i+1} can be written as follows:

$$H_{i+1} = H_1 - \overline{A}^i(\overline{W}^i)^T . \tag{4.1}$$

Proof
We proceed by induction. For $i = 1$, (4.1) is obviously true. Assume now the validity of (4.1) up to the index $i - 1$, say:

$$H_i = H_1 - \overline{A}^{i-1}(\overline{W}^{i-1})^T . \tag{4.2}$$

The induction is completed by substituting (4.2) in the first term of the right-hand side of (3.5) and noting that $\overline{A}^i(\overline{W}^i)^T = \overline{A}^{i-1}(\overline{W}^{i-1})^T + H_i a_i w_i^T H_i$. Q.E.D.

Theorem 4.2
Let A^i and W^i be defined by (3.23) and (3.25), with A^i full rank and w_i satisfying (3.6). Let H_1 be nonsingular. Define the matrix $Q^i \in R^{i,i}$ by

$$Q^i = (W^i)^T H_1 A^i \ . \tag{4.3}$$

Then Q^i is strongly nonsingular.

Proof
Multiplying (4.1) on the right by A^i and using (3.27) and (3.34) gives

$$H_1 A^i = \overline{A}^i R_1 \ . \tag{4.4}$$

Multiplying (4.4) on the left by $(W^i)^T$ and using (3.35) and (4.3) gives

$$Q^i = R_2^T R_1 \ , \tag{4.5}$$

which proves the thesis, since Q^i is decomposed into the product of a lower by an upper triangular nonsingular matrix. Q.E.D.

Theorem 4.3
The matrix H_{i+1} can be written in the form

$$H_{i+1} = H_1 - H_1 A^i [(W^i)^T H_1 A^i]^{-1} (W^i)^T H_1 \ . \tag{4.6}$$

Proof
From (4.4) we get

$$\overline{A}^i = H_1 A^i R_1^{-1} \ . \tag{4.7}$$

Multiplying (4.1) on the left by $(W^i)^T$ and using (3.28) and (3.35), we get $(W^i)^T H_1 = R_2^T (\overline{W}^i)^T$ so that

$$(\overline{W}^i)^T = (R_2^T)^{-1} (W^i)^T H_1 \ . \tag{4.8}$$

Substituting relations (4.7) and (4.8) in Theorem 4.1 gives $H_{i+1} = H_1 - H_1 A^i (R_2^T R_1)^{-1} (W^i)^T H_1$ and the thesis follows by using (4.5). Q.E.D.

Equation (4.6) has been used by Spedicato (1987c), Bodon (1988), Deng and Spedicato (1988) and Yang (1988c) in the determination of w_i by an optimal conditioning criterion of the kind used in quasi-Newton methods (Spedicato and Oren 1976, Spedicato and Greenstadt 1978). The optimal w_i is obtained by minimizing a strict bound (Spedicato 1987b and Spedicato and Burmeister 1988) of the condition number of certain bordered positive matrices. If the bound is applied at every iteration to the matrix $(Q^i)^T Q^i$, Q^i defined by (4.3), and $H_2 = I$, then it is proved (Deng and Spedicato 1988) that the optimal w_i satisfies the linear system

$$(A^{i-1})^T w_i = 0 \; , \tag{4.9}$$

$$a_i^T w_i = 1 \; . \tag{4.10}$$

Using Theorems 3.3 and 3.4, the general solution of the homogeneous system (4.9) can be put in the form $w_i = H_i^T s$, $s \in R^n$ and arbitrary. Substituting in (4.10), we get $s^T H_i a_i = 1$, which is an equation of the same form as (3.6) having the following minimal Euclidean norm solution:

$$s_i^+ = \frac{H_i a_i}{(H_i a_i)^T (H_i a_i)} \; . \tag{4.11}$$

It is easy to show by induction that the choice $w_i = H_i^T s^+$ generates symmetric matrices, implying also that $w_i = s_i^+$. The resulting sequence H_i is identical with that generated by the modified Huang algorithm described in Chapter 5.

4.3   REPRESENTATION OF H_{i+1} IN TERMS OF A AND H_1

In this section we derive a modified form of equation (4.6) which does not require matrix inversion and we apply it to the computation of the search vector. For this purpose, we introduce a representation of the matrix W^i in terms of A^i and we discuss some of its properties.

Theorem 4.4
Let A be n by n and nonsingular. Let H_1 be nonsingular and let the sequence w_i satisfy (3.6) and be given by

$$w_i = \sum_{j=1}^n \beta_{i,j} a_j \tag{4.12}$$

or in matrix terms

$$W^n = A^n L \; , \tag{4.13}$$

where L is lower triangular. Then in (4.11) the coefficient $\beta_{i,i}$ is nonzero, so that L is nonsingular.

Proof
Observe that (4.12) implies that $W^i = A^n L^i$, where L^i is the matrix comprising the first i columns of L. Using Theorem 4.2, it follows that $Q^i = (L^i)^T A^n H_1 A^i$ is nonsingular. Thus L^i is full rank for every i, implying that L is nonsingular and that the coefficients $\beta_{i,i}$ are nonzero. Q.E.D.

Theorem 4.5
Consider the matrices H_i generated by update (3.5) where H_1 is nonsingular and w_i satisfies the conditions of Theorem 4.4. Then for $i = 0, \ldots, n-1$, the vectors $H_{i+1}a_{i+1}, \ldots, H_{i+1}a_n$ are linearly independent.

Proof
For $i = 0$, the theorem follows from the nonsingularity of H_1 and A. For $i > 0$, we prove it by contradiction. If the theorem were not true, there would exist a nonzero vector $s \in R^{n-i}$ such that

$$H_{i+1}{}^T A^{n-i} s = 0 . \tag{4.14}$$

The above relation states that the vector $A^{n-i}s$ belongs to $\text{Null}(H_{i+1}{}^T)$. Since $\text{Null}(H_{i+1}{}^T)$ is spanned by the columns of W_i we can write, using (4.13), that

$$A^{n-i}s = A^n L^i u \tag{4.15}$$

for some $u \in R^i$. Decompose L^i as follows:

$$L^i = \begin{bmatrix} L^{i1} \\ L^{i2} \end{bmatrix} , \tag{4.16}$$

with L^{i1} triangular nonsingular of order i. Since $A^n = (A^i, A^{n-i})$, (4.15) can be written as

$$A^i L^{i1} u + A^{n-i}(L^{i2}u - s) = 0 . \tag{4.17}$$

Since the columns of A are linearly independent, it follows that $L^{i1}u = 0$ and $L^{i2}u - s = 0$. From the nonsingularity of L^{i1}, it follows that both u and s are zero. Since s was assumed to be nonzero, a contradiction has been established, proving the theorem. Q.E.D.

The proof of the following theorem is omitted since it is the reformulation of the previous theorem in terms of the transpose.

Theorem 4.6
Consider update (3.5) with H_1 nonsingular and w_i of the form (4.11) and satisfying (3.6). Then, for $i = 0, \ldots, n-1$, the vectors $H_{i+1}w_{i+1}, \ldots, H_{i+1}w_n$ are linearly independent.

The lemma that is now given is used in the following and in other theorems further on.

Lemma 4.1
Let B be a nonsingular i by i matrix. Let u and v be vectors in R^i. Let σ be a scalar. If the matrix B' is given by

$$B' = \begin{bmatrix} B & v \\ u^T & \sigma \end{bmatrix},$$

(4.18)

then B' is nonsingular if and only if

$$\sigma - u^T B^{-1} v \neq 0 .$$

(4.19)

If B is positive definite and $u = v$, then B' is positive definite if and only if

$$\sigma > u^T B^{-1} u .$$

(4.20)

Proof
We can write B' as follows:

$$B' = \begin{bmatrix} B & 0 \\ 0 & 1 \end{bmatrix} \begin{bmatrix} I_i & B^{-1} v \\ u^T & \sigma \end{bmatrix} .$$

(4.21)

Using Laplace expansion, we get

$$\det(B') = \det(B) (\sigma - u^T D^{-1} v)$$

(4.22)

and the theorem follows immediately. Q.E.D.

Theorem 4.7
Consider update (3.5) with H_1 the unit matrix and w_i satisfying (3.6). Let W^n have the form (4.13) with L^n triangular nonsingular. Then, for $i = 1, \ldots, n$, H_{i+1} has the form

$$H_{i+1} = I - A^i B^i (L^i)^T (A^n)^T ,$$

(4.23)

where B_i is the following strongly nonsingular i by i matrix:

$$B^i = [(L^i)^T A A^i]^{-1} .$$

(4.24)

Moreover B^i is updated as follows: $\overline{B}^0 = 0$ and

$$B^i = \overline{B}^{i-1} + uv^T , \tag{4.25}$$

where $\overline{B}^{i-1}$ consists of B^{i-1} bordered with an extra null row and column and u, v are vectors in R^i given by

$$u = \mu \begin{bmatrix} B^{i-1}q \\ -1 \end{bmatrix} , \tag{4.26}$$

$$v = \tau \begin{bmatrix} (B^{i-1})^T r \\ -1 \end{bmatrix} , \tag{4.27}$$

where

$$q = (W^i)^T a_i , \tag{4.28}$$

$$r = (A^i)^T w_i , \tag{4.29}$$

with μ, τ arbitrary nonzero scalars such that

$$\mu\tau = \frac{1}{\sigma - r^T B^{i-1} q} , \tag{4.30}$$

where σ is the scalar given by

$$\sigma = w_i^T a_i \tag{4.31}$$

and the denominator in (4.30) is nonzero.

Proof
Equations (4.23) and (4.24) are immediately obtained by substituting (4.13) in (4.6), with the implication that B^i is strongly nonsingular. By definition, $B^i = [(W^i)^T A^i]^{-1}$ or, in structured form,

$$(B^i)^{-1} = \begin{bmatrix} (W^{i-1})^T A^{i-1} & (W^{i-1})^T a_i \\ w_i^T A^{i-1} & w_i^T a_i \end{bmatrix} \tag{4.32}$$

or, with the given definitions,

$$(B^i)^{-1} = \begin{bmatrix} (B^{i-1})^{-1} & q \\ r^{\mathrm{T}} & \sigma \end{bmatrix} . \tag{4.33}$$

Since B^i and B^{i-1} are nonsingular, it follows from Lemma 4.1 that $\sigma - r^{\mathrm{T}}B^{i-1}q \neq 0$; hence the denominator in (4.30) is nonzero. Update (4.25) with the definitions (4.26)–(4.31) follows after some algebra by verifying the identity $(\bar{B}^{i-1} + uv^{\mathrm{T}})B^i = I_i$. Q.E.D.

The following theorem expresses the search vector in terms of the rows of A.

Theorem 4.8
Let H_1 be nonsingular and $m = n$. Suppose that the vector z_i satisfies for every i condition (3.2). Let W^n have the form

$$W^n = A^n L , \tag{4.34}$$

with L a banded lower triangular nonsingular matrix whose band width is $r + 1$, and $r \leqslant n - 1$ is an assigned integer. Then for $p = \min(n, i + r)$ and some coefficients β_j the following relation holds:

$$p_{i+1} = z_{i+1} - \sum_{j=1}^{p} \beta_j a_j. \tag{4.35}$$

Proof
From (3.1) and (4.23) we get

$$p_{i+1} = z_{i+1} - A^n L^i (B^i)^{\mathrm{T}} (A^i)^{\mathrm{T}} z_{i+1} , \tag{4.36}$$

where L^i is the matrix comprising the first i columns of L. Since only the first p rows of L^i are nonzero, only the first p elements of the vector $L^i(B^i)^{\mathrm{T}}(A^i)^{\mathrm{T}}z_{i+1}$ are nonzero; hence (4.35) follows immediately. Q.E.D.

4.4 REPRESENTATION OF p_i IN TERMS OF $2i$ VECTORS

The following theorem shows that the search vector p_i can be computed without explicit use of the matrix H_i, provided that $2i - 2$ vectors are stored and z_i, w_i are given.

Theorem 4.9
Equation (3.1) for the search vector p_i is equivalent for $i > 1$ to the following formula:

$$p_i = H_1^{\mathrm{T}}z_i - \sum_{j=1}^{i-1} u_j s_j^{\mathrm{T}} z_i , \tag{4.37}$$

where s_j, u_j are vectors in R^n given by $s_1 = H_1 a_1$, $u_1 = H_1^T w_1$ and, for $i > 1$,

$$s_i = H_1 a_i - \sum_{j=1}^{i-1} s_j u_j^T a_i , \qquad (4.38)$$

$$u_i = H_1^T w_i - \sum_{j=1}^{i-1} u_j s_j^T w_i . \qquad (4.39)$$

Proof

If we define $s_i = H_i a_i$ and $u_i = H_i^T w_i$, equation (4.1) gives

$$H_i = H_1 - \sum_{j=1}^{i-1} s_j u_j^T , \qquad (4.40)$$

from which the relations of the theorem follow immediately. Q.E.D.

Remark 4.2

The computation of p_i by (4.37) requires the storage of the $2i - 2$ vectors s_j, u_j. There is therefore a saving in the memory requirement when the system is underdetermined, with fewer than $n/2$ equations.

Remark 4.3

When $H_1 = I$ the computation of p_i by use of (4.37)–(4.39) needs fewer multiplications generally than through the use of the matrix H_i. Indeed, at the ith step, $3n(i-1)$ multiplications are required for the scalar products $s_j^T z_i$, $s_j^T w_i$, $u_j^T a_i$; $3n(i-1)$ multiplications are also required for the linear combinations, leading to a total count of $3n^3 + O(n^2)$ multiplications, versus $4n^3 + O(n^2)$ via the matrix H_i, in the case of determined systems. The advantage is even greater in the case of underdetermined systems, the total count being $3nm^2$ versus $4n^2m$.

We can use relations (4.37)–(4.39) to give an equivalent formulation of ALGORITHM 1.

ALGORITHM 2: The ABS algorithm using $2i$ vectors

(A2) Let $x_1 \in R^n$ be arbitrary. Let $H_1 \in R^{n,n}$ be arbitrary nonsingular. Assign matrices $S = (s_1, \ldots, s_m)$ and $U = (u_1, \ldots, u_m)$ in $R^{n,m}$ to be used for storing vectors. Let $p_1 \in R^n$ be such that $p_1^T a_1 \neq 0$ and let $u_1 \in R^n$ be such that $u_1^T a_1 = 1$. Compute $s_1 = H_1 a_1$. Set $i = 1$ and $iflag = 0$. Go to (E2).

(B2) Compute the scalar $\tau_i = a_i^T x_i - b^T e_i$ and the vector s_i, the ith column of S, by (4.38).

(C2) If $s_i \neq 0$, go to (D2). If $s_i = 0$ and $\tau_i = 0$, set $x_{i+1} = x_i$, $u_i = 0$, iflag $=$ iflag $+ 1$ and go to (G2) if $i < m$; otherwise stop. x_{m+1} solves the system. If $s_i = 0$ and $\tau_i \neq 0$, set iflag $= -i$ and stop.

(D2) Compute the search vector p_i by (4.37) with $z_i \in R^n$ any vector such that $a_i^T p_i \neq 0$.

(E2) Compute the new approximation to the solution x_{i+1} with (3.3) and (3.4). If $i = m$, stop x_{m+1} solves the given system. If $i = 1$, go to (G2); otherwise go to (F2).

(F2) Compute the vector u_i, the ith column of U, by (4.39).

(G2) Increment the index i by one and go to (B2).

4.5   REPRESENTATION OF p_i IN TERMS OF $n - i$ VECTORS

In this section we give another representation of the algorithm which does not make explicit use of the matrix H_i. This representation essentially updates a basis of H_i^T, which is all that is needed in the construction of the search vector p_i. The representation is obtained for the subclass of the ABS class where the vectors z_i and w_i are multiple of each other. Note that, from Theorem 3.12, there is no loss of generality in making this restriction.

Theorem 4.10
Let $z_1, \ldots, z_m$ be a set of admissible vectors in step (D1) of ALGORITHM 1. Define in step (F1) w_i by

$$w_i = \frac{z_i}{z_i^T H_i a_i} \; . \tag{4.41}$$

Consider for $i = 1, \ldots, m$ the $m - i$ vectors $u_j^{i+1} \in R^n$ defined by

$$u_j^{i+1} = u_j^i - \frac{a_i^T u_j^i}{(a_i^T u_i^i)} u_i^i \; , \qquad j = i+1, \ldots, m \; , \tag{4.42}$$

with

$$u_j^1 = H_1^T z_j \; , \qquad j = 1, \ldots, m \; . \tag{4.43}$$

Then the following identity holds for $j = i, \ldots, m$:

$$H_i^T z_j = u_j^i \; . \tag{4.44}$$

Proof
We proceed by induction. For $i = 1$ the result is immediate. Assuming the validity of (4.44) up to i, we observe that for the given choice of w_i it holds for $j = i + 1, \ldots, m$ that

$$H_{i+1}^T z_j = H_i^T z_j - \frac{a_i^T H_i^T z_j}{a_i^T H_i^T z_i} H_i^T z_i \qquad (4.45)$$

or, by the induction assumption,

$$H_{i+1}^T z_j = u_j^i - \frac{a_i^T u_j^i}{a_i^T u_i^i} u_i^i , \qquad j = i + 1, \ldots, m . \qquad (4.46)$$

From (4.42) and (4.46) the identity $H_{i+1}^T z_j = u_j^{i+1}$ is established, completing the induction. Q.E.D.

Corollary 4.1
The vectors $u_j^i, j = i, \ldots, m$ are linearly independent. If $m = n$ they form a basis of H_i^T.

Corollary 4.2
Let $m = n$; then the linear variety containing all solutions of the first i equations of the given system is the set of vectors $\bar{x}$ of the form

$$\bar{x} = x_{i+1} + U^i d , \qquad (4.47)$$

where x_{i+1} is the approximation of the solution obtained at the ith step of the ABS algorithm, $U^i = (u_{i+1}^{i+1}, \ldots, u_n^{i+1})$ is defined by (4.42) and $d \in R^{n-i}$ is arbitrary.

Corollary 4.3
If w_i is given by (4.41), the search vector p_i is equal to u_i^i.

An apparent problem in Theorem 4.10 is that the vectors $z_1, \ldots, z_m$ must be admissible, say they must satisfy condition (3.2). Since (3.2) involves H_i, a direct testing of the admissibility may not be possible in advance. Since, however, choices of $z_1, \ldots, z_m$ exist, which are independent of H_i and whose admissibility can be guaranteed *a priori* (see Remark 4.4), the theorem is of practical use.

Since the recursions of Theorem 4.10 essentially construct a basis of H_i^T and p_i is in general an arbitrary vector in the range of H_i^T, subject to nonorthogonality with a_i, it is possible to exploit them for a reformulation of the ABS algorithm. In order to do that, we introduce the following.

Definition 4.1

A set of m linearly independent vector $z_1, \ldots, z_m$ is called H_i-admissible if, given H_1 nonsingular and $a_1, \ldots, a_m$ linearly independent vectors, $z_i^T H_i a_i$ is nonzero when H_i is constructed according to (3.5) and (4.41) for $i = 1, \ldots, m$.

Remark 4.4

It follows from Theorem 3.7 that the vectors $a_1, \ldots, a_m$ are H_i-admissible. It will be proved in Chapter 6 that the vectors $e_1, \ldots, e_n$ are H_i-admissible, provided that A^n is strongly nonsingular.

Remark 4.5

If the recursion (4.41) is started with $z_1, \ldots, z_n$ being H_i-admissible, then the scalar products $a_i^T u_i^i$ are all nonzero if the vectors a_i are linearly independent. From Theorem 3.7, $a_j^T u_i^i$ is zero for $j < i$. Thus $a_i^T u_i^i$ is zero if and only if a_i is a linear combination of $a_1, \ldots, a_{i-1}$.

From the previous considerations it is clear that the following algorithm is a reformulation of ALGORITHM 1.

ALGORITHM 3: The ABS algorithm using $u - i$ vectors

(A3) Let $x_1 \in R^n$ be arbitrary. Let $H_1 \in R^{n,n}$ be arbitrary positive definite. Let $z_1, \ldots, z_m \in R^n$ be H_i-admissible vectors. Define the m vectors $u_1^1, \ldots, u_m^1$ by (4.43). Set $i = 1$ and $iflag = 0$.

(B3) Compute the scalars $\tau_i = a_i^T x_i - b^T e_i$ and $\delta_i = a_i^T u_i^i$.

(C3) If $\delta_i \neq 0$, go to (D3); if $\delta_i = 0$ and $\tau_i = 0$, set $x_{i+1} = x_i$, $iflag = iflag + 1$, $u_j^{i+1} = u_j^i$ for $j = i+1, \ldots, m$ and go to (B3) if $i < m$, otherwise stop, x_{m+1} solves the system. If $\delta_i = 0$ and $\tau_i \neq 0$ set $iflag = -i$ and stop.

(D3) Compute the search vector p_i by

$$p_i = U^i d_i , \tag{4.48}$$

where $U^i = (u_i^i, \ldots, u_m^i)$ and $d_i \in R^{m-i-1}$ is arbitrary save that $p_i^T a_i \neq 0$.

(E3) Compute the new approximation x_{i+1} with equations (3.3) and (3.4). If $i = m$, stop, x_{m+1} solves the system.

(F3) Update the vectors u_j^i by (4.42) for $j = i+1, \ldots, m$.

(G3) Increment the index i by one and go to (B3).

Remark 4.6

With respect to ALGORITHM 1, in ALGORITHM 3, if $m < n$, only a subset of the basis of H_i^T is built by recursion (4.42); hence the set of vectors generated by ALGORITHM 3 is only a subset of the set that can be generated by ALGORITHM 1.

Remark 4.7

The number of multiplications required by ALGORITHM 3 at the ith step is of order $3(m-i)n$ and thus for m steps $(3/2)nm^2$, say $(3/2)n^3$ for a determined system. Thus there is a decrease in the computational cost compared with ALGORITHM 1, where the numbers are respectively $4n^2m$ and $4n^3$. If $p_i = u_i^i$, the numbers are further reduced to nm^2 and n^3. The memory occupation is nm if the choice $H_1 = I$ is made; otherwise it is $n^2 + nm$.

Remark 4.8

ALGORITHM 3 is equivalent to a class of algorithms proposed by Sloboda (1978) for linear systems, see Abaffy (1988b). Sloboda established that the algorithm is well defined in the case $z_i = e_i$, which is considered in detail in Chapter 6.

4.6 REPRESENTATION OF p_i IN TERMS OF i VECTORS

As discussed in Chapter 2, the basic requirement on z_i for forcing finite termination is equation (2.10), which we restate in term of the search vector p_i:

$$a_j^{\mathrm{T}} p_i = 0 , \qquad j = 1, \ldots, i-1 . \tag{4.49}$$

Equations (4.49) constitute a homogeneous system for the unknown p_i, which can be solved by the ABS method itself. If we take $H_1 = I$, z_i as the starting point p_i^1, with the z_i H_i-admissible vectors and we use for the first $i-1$ equations of the ith system the search vectors used in the $(i-1)$th system, then the ABS iteration for p_i has the form

$$p_i^{j+1} = p_i^j - \frac{a_j^{\mathrm{T}} p_i^j}{a_j^{\mathrm{T}} p_j} p_j , \qquad j = 1, \ldots, i-1 \tag{4.50}$$

implying that for p_i

$$p_i = p_i^i . \tag{4.51}$$

Equation (4.50) requires the storage of the i vectors p_j. The number of multiplications required to compute the sequence p_i is the same as in ALGORITHM 3, i.e. $(3/2)nm^2$, against $4n^2m$ in the standard approach.

4.7 BIBLIOGRAPHICAL REMARKS

Theorems 4.1—4.8 are due to Broyden and have appeared in the Abaffy *et al.* (1982, 1984a,b); some of them are given here in a slightly more general form. Note that equation (4.6) has appeared also in a posthumous paper of Egervary (1960). Theorem 4.9 is new and due to Spedicato. Theorem 4.10 is due to Abaffy and has appeared in the context of the scaled ABS algorithm in the paper by Abaffy and Spedicato (1985). A different derivation of ALGORITHM 3 is presented by Gu (1988). The representation for p_i given in section 4.6 is due to Bodon (1989).

5

The Huang and the modified Huang algorithm

5.1 INTRODUCTION

In this chapter we begin the study of particular parameter choices in the ABS class by considering, in a more general form, the algorithm that Huang proposed in 1975 and that motivated Abaffy in 1979 to develop the first formulation of the ABS class.

The algorithm will be named in different ways. The name 'symmetric algorithm' emphasizes the use of the symmetric update for the matrix H_i. The name 'Huang' acknowledges the original Huang algorithm. The name 'implicit Gram–Schmidt' recognizes that the search vectors are identical with those generated by the classical Gram–Schmidt process, so that the algorithm can be considered as an alternative formulation of such a fundamental procedure. The name 'implicit LQ' refers to the property that the coefficient matrix is implicitly factorized into the product of a lower triangular and an orthogonal matrix.

The algorithm is related to other methods in the literature. In particular it is equivalent to the Brent (1973) method for nonlinear systems, when restricted to linear systems. It can also be used for computing the pseudoinverse and we discuss its connection with the Pyle (1964) method.

5.2 DEFINITION AND BASIC PROPERTIES OF THE HUANG ALGORITHM

Definition 5.1

The symmetric algorithm is the algorithm corresponding to the following parameter choices in the ABS class: H_1 symmetric positive definite, $z_i = a_i$, $w_i = a_i/a_i^T H_i a_i$.

Remark 5.1

From Theorem 3.7 the symmetric algorithm is well defined, the sequence H_i consists of symmetric matrices and the denominator $a_i^T H_i a_i$ is strictly positive.

Definition 5.2
The Huang algorithm (or the implicit Gram–Schmidt or the implicit LQ algorithm) is the algorithm corresponding to the choice $H_1 = I$ in the symmetric algorithm.
We now establish some properties of these algorithms.

Theorem 5.1
The search vectors p_i generated by the symmetric algorithm are H_1^{-1}-conjugate.

Proof
Consider the matrices $\overline{A}^i$ and $\overline{W}^i$ defined in (3.32) and (3.33). Since for the symmetric algorithm $\overline{A}^i = (p_1, \ldots, p_i)$, $\overline{W}^i = (p_1/a_1^T H_1 a_1, \ldots, p_i/a_i^T H_i a_i)$ the theorem follows from statement (c) of Theorem 3.10. Q.E.D.

Corollary 5.1
The search vectors generated by the Huang algorithm are orthogonal.

Remark 5.2
Define the matrix Q^i by

$$Q^i = P^i D^i , \tag{5.1}$$

with D^i the i by i diagonal matrix with diagonal elements:

$$D^i_{j,j} = (p_j^T p_j)^{-1/2} , \qquad j = 1, \ldots, i . \tag{5.2}$$

For the Huang algorithm, Corollary 5.1 implies that Q^i has orthogonal columns. Thus, for $m = n$, (3.39) takes the form

$$A = \overline{L} \, \overline{Q} , \tag{5.3}$$

with $\overline{Q} = D^{-n} P^{-1}$ orthogonal and with $\overline{L} = L D^n$ a lower triangular matrix, thereby justifying the name 'implicit LQ' algorithm.

The relation with the classical Gram–Schmidt algorithm is given by the following.

Theorem 5.2
Let $m = n$. Then the search vectors generated by the Huang algorithm are identical with the vectors generated by the classical unstabilized Gram–Schmidt method when applied to the vectors $a_1, \ldots, a_n$, i.e.

$$p_{i+1} = a_{i+1} - \sum_{j=1}^{i} \frac{p_j^T a_{i+1}}{p_j^T p_j} p_j . \tag{5.4}$$

Proof
Applying Theorem 4.8 to the Huang algorithm, we have

$$p_{i+1} = a_{i+1} - \sum_{j=1}^{i} \beta_j a_j \qquad (5.5)$$

or

$$P^{i+1} = A^{i+1} U^{i+1} \qquad (5.6)$$

for some unit upper triangular matrix U^{i+1}. Thus $A^{i+1} = P^{i+1}(U^{i+1})^{-1}$ with $(U^{i+1})^{-1}$ again unit upper triangular, implying that, for $j = 1, \ldots, i+1$,

$$a_j = p_j + \sum_{k=1}^{j-1} \tau_{k,j} p_k, \qquad (5.7)$$

with $\tau_{k,j} = e_k^{\mathrm{T}}(U^{i+1})^{-1} e_j$. Substituting (5.7) in (5.5) yields

$$p_{i+1} = a_{i+1} - \sum_{j=1}^{i} \mu_{j,i+1} p_j \qquad (5.8)$$

for some scalar $\mu_{j,i+1}$. Multiplying (5.8) on the left by p_k, $k \leq i$, and using the orthogonality of the search vectors, we get

$$\mu_{k,i+1} p_k^{\mathrm{T}} p_k = p_k^{\mathrm{T}} a_{i+1} . \qquad (5.9)$$

Hence (5.8) takes the form (5.4). Q.E.D.

The next theorem shows that for any x_1 the iterates x_i, x_{i+1} generated by the Huang algorithm are as close as possible in Euclidean norm. The following two theorems show that, if some additional mild condition is imposed on x_1, the iterates x_i have the smallest possible Euclidean norm. These results can be viewed as establishing remarkable regularization properties of the Huang algorithm. The theorems can be extended to the symmetric algorithm by using the weighted Euclidean norm $\|x\| = x^{\mathrm{T}} H_1^{-1} x$.

Theorem 5.3
Consider the sequence x_i generated by the Huang algorithm. Then for $i = 1, \ldots, m$ the following relation holds

$$x_{i+1} = \arg\left(\min_{x'} \|x' - x_i\|\right), \qquad x' \in \Omega^{n-i}, \tag{5.10}$$

where Ω^{n-i} is the linear variety containing all solutions of the first i equations.

Proof
By Theorem 3.15 any vector in Ω^{n-1} can be written in the form $x' = x_{i+1} + H_{i+1}{}^T s$ for some $s \in R^n$. Then $x' - x_i = x_{i+1} - x_i{}^T H_{i+1}{}^T s = u_i + H_{i+1}{}^T s$, with $u_i = x_{i+1} - x_i$. Hence, from symmetry of H_{i+1},

$$\|x' - x_i\|^2 = \|u_i\|^2 + \|H_{i+1}{}^T s\|^2 + 2u_i{}^T H_{i+1} s. \tag{5.11}$$

Since $u_i = \alpha_i H_i a_i$, we have $u_i{}^T H_{i+1} s = -\alpha_i a_i{}^T H_i H_{i+1} s = -\alpha_i a_i{}^T H_{i+1} s = 0$, from Theorems 3.6 and 3.3, implying (5.10). Q.E.D.

Theorem 5.4
Consider the Huang algorithm with the following choice for x_1:

$$x_1 = \sum_{j=1}^{k} \beta_j a_j, \tag{5.12}$$

with $k < m$ and the β_j some scalars. Then, for $i \geq k$, x_{i+1} is the vector in Ω^{n-1} with minimal Euclidean norm.

Proof
Let x_{i+1} be the iterate generated by the Huang algorithm at the ith iteration. Then from Theorem 3.15 any vector in Ω^{n-1} can be written in the form $\bar{x} = x_{i+1} + H_{i+1}{}^T s$ for some $s \in R^n$. Noting that x_{i+1} has the form

$$x_{i+1} = \sum_{j=1}^{k} \beta_j a_j + \sum_{j=1}^{i} \alpha_j p_j \tag{5.13}$$

with α_j as in (3.4), we have

$$\bar{x}^T \bar{x} = x_{i+1}{}^T x_{i+1} + s^T H_{i+1} H_{i+1}{}^T s + 2\left(\sum_{j=1}^{k} \beta_j a_j{}^T H_{i+1}{}^T s\right)$$

$$+ \sum_{j=1}^{i} \alpha_j a_j{}^T H_j H_{i+1}{}^T s \Bigg) \ . \tag{5.14}$$

From Theorem 3.6 and symmetry of H_{i+1}, $H_j H_{i+1}{}^T = H_{i+1}$; hence by Theorem 3.3 all terms in the summations are zero. The statement of the theorem follows immediately. Q.E.D.

If $x_1 = \beta a_1$ the previous theorem implies that the whole sequence of iterates $x_2, \ldots, x_{m+1}$ generated by the Huang algorithm consists of minimal Euclidean norm solutions of the associated subsystems. The next theorem shows that the condition $x_1 = \beta a_1$ is also necessary for the whole sequence of solutions to be of minimal Euclidean norm.

Theorem 5.5
The vector x_2 generated by the Huang algorithm is the minimal Euclidean norm vector in Ω^{n-1} if and only if $x_1 = \beta a_1$ for some arbitrary scalar β.

Proof
The minimal Euclidean norm solution of the equation $a_1{}^T x = e_1{}^T b$ has the form

$$\bar{x} = a_1{}^+ e_1{}^T b \ , \tag{5.15}$$

with $a_1{}^+$ the Moore–Penrose pseudoinverse of $a_1{}^T$. Since for $a_1 \neq 0$ we have $a_1{}^+ = a_1/a_1{}^T a_1$, (5.15) gives

$$\bar{x} = \frac{e_1{}^T b}{a_1{}^T a_1} a_1 \ . \tag{5.16}$$

For arbitrary x_1, the vector x_2 generated by the Huang algorithm has the form $x_2 = (e_1{}^T b/a_1{}^T a_1)a_1 + (a_1{}^T a_1 x_1 - a_1{}^T x_1 a_1)/a_1{}^T a_1$; hence $\bar{x} = x_2$ if and only if $x_1 = \beta a_1$ for arbitrary β. Q.E.D.

Remark 5.3
For arbitrary starting point x_1, one can show (Yang 1988b) that x_{i+1} has the form $x_{i+1} = \bar{x}_i + H_{i+1} x_1$, where $\bar{x}_i$ is the solution of the first i equations of minimal Euclidean norm.

Theorem 5.6
Let $x_1 = \beta a_1$, with β an arbitrary scalar. Then the iterates x_i generated by the Huang algorithm satisfy the following inequality for $2 \leq i \leq m$:

$$\|x_i\| \leq \|x_{i+1}\| \ , \tag{5.17}$$

$$\|x_{i+1} - x^+\| \leq \|x_i - x^+\| \ . \tag{5.18}$$

Proof
Just use the orthogonality of p_i and the identity

$$x^+ = (\beta - \alpha_1)p_1 - \alpha_2 p_2 - \ldots - \alpha_m p_m. \text{ Q.E.D.}$$

It was observed in Remark 3.3 that, in exact arithmetic, symmetric matrices can be generated by an infinite number of choices of w_i. Similarly, infinite choices exist for z_i which generate the same search vector that is given by the Huang algorithm. For instance, instead of $z_i = a_i$ one can take

$$z_i = S_i a_i , \tag{5.19}$$

with S_i defined by (3.20). The simplest choice for S_i in (3.20) is $S_i = H_i = H_i^T$, implying that

$$z_i = H_i a_i \tag{5.20}$$

and so

$$p_i = H_i^T(H_i a_i) , \tag{5.21}$$

$$\alpha_i = \frac{e_i^T r(x_i)}{z_i^T z_i} \tag{5.22}$$

With the choice $z_i = a_i$, $w_i = a_i/a_i^T H_i a_i$ the update of H_i can be written as follows:

$$H_{i+1} = H_i - \frac{p_i p_i^T}{a_i^T p_i} \tag{5.23}$$

or also, since $a_i^T p_i = a_i^T H_i a_i = (H_i^T a_i)^T(H_i^T a_i) = p_i^T p_i$,

$$H_{i+1} = H_i - \frac{p_i p_i^T}{p_i^T p_i} . \tag{5.24}$$

When choice (5.20) for z_i is made, a natural choice for w_i is $w_i = z_i/z_i^T H_i a_i$ or

$$w_i = \frac{z_i}{z_i^T z_i} \tag{5.25}$$

which leads to the following formula for the update:

$$H_{i+1} = H_i - \frac{z_i p_i^T}{z_i^T z_i} . \tag{5.26}$$

However, (5.23) or (5.24) can also be used when z_i is defined by (5.20), since we have the identity $z_i = p_i$.

Extensive numerical experiments, presented in Chapter 10, show that choices (5.20) and (5.25) generate an algorithm which is more stable numerically. To recognize this fact, we denote this algorithm by a special name.

Definition 5.3
The version of the Huang algorithm where z_i is given by (5.20) and w_i by (5.25) is called the modified Huang algorithm.

Remark 5.4
The property that $a_i^T H_i a_i$ is strictly positive (see Theorem 3.7) may be violated owing to round-off errors. Thus, if $z_i = a_i$, the denominators in (3.4) and (3.5) can become negative. Such an occurrence cannot happen if we use (5.22), (5.26) or (5.24).

Remark 5.5
The vector p_i is defined in the Huang algorithm as the projection of a_i on Range(H_i). In the modified Huang algorithm it is defined as the reprojection on Range(H_i) of the vector defined by the Huang algorithm. Such a reprojection clearly tends to annihilate components in the orthogonal complement to Range(H_i), say in A^{i-1}, which may have been introduced by round-off errors.

Remark 5.6
With choice (5.25) the vector w_i is the vector of minimal Euclidean norm which satisfies the projection condition (3.6). It is also the vector of minimal Euclidean norm which satisfies the optimal conditioning equations (4.9) and (4.10).

Remark 5.7
Owing to round-off errors the matrix L_i in the implicit factorization (3.38) may have nonzero elements above the main diagonal, implying a violation of the condition $a_j^T p_i = 0, j < i$, which is essential for the termination of the algorithm. Yang (1988d) gives arguments to suggest that, if the above diagonal elements are of order ε for the Huang algorithm, they are of order ε^2 for the modified Huang algorithm.

Remark 5.8
In exact arithmetic the eigenvalues of H_{i+1} are zero, with multiplicity i, and one, with multiplicity $u - i$. Let ε be the size of the interval into which the zero eigenvalues spread owing to round-off errors when the Huang update is used. Then, if the modified Huang update is used, the corresponding size of the interval is ε^2, see Broyden (1989).

5.3   ALTERNATIVE REPRESENTATION OF THE HUANG AND THE MODIFIED HUANG ALGORITHMS

In this section we apply the alternative formulations of the ABS class of Chapter 3, to the Huang and the modified Huang algorithms. Some of these formulations are convenient in terms of memory requirement and/or computational cost. The memory requirement of both the Huang and the modified Huang algorithms, as previously formulated, is no more than $n^2/2 + 0(n)$ positions, if symmetry of H_i is taken into account. The computational cost for the Huang algorithm is no more than

$(3/2)n^2m + O(nm)$ multiplications, for the modified Huang algorithm it is no more than $(5/2)n^2m + O(nm)$ (again having taken into account the symmetry of H_i).

If the Gram–Schmidt relation (5.4) is used to define the search vectors of the Huang algorithm, then the memory requirement is no more than $nm + O(n)$ positions (disregarding the possibility of overwriting A^i by P^i) and the computational cost is no more than $nm^2 + O(nm)$.

Let us consider the representation presented in sections 4.2 and 4.6. In Huang algorithm, $W^i = A^iD^i$, with D^i the diagonal nonsingular matrix whose diagonal elements are $1/a_i^TH_ia_i$. Therefore, from (4.6), we get

$$H_{i+1} = I - A^i(A^{iT}A^i)^{-1}A^{iT} \tag{5.27}$$

or letting $G^i = (A^{iT}A^i)^{-1}$,

$$H_{i+1} = I - A^iG^iA^{iT} . \tag{5.28}$$

G^i is a positive definite matrix whose update can be obtained along similar lines as for the update of B^i in section 4.3. Indeed we have from the definition of G_i

$$G^{i+1} = \begin{bmatrix} (G^i)^{-1} & A^{iT}a_{i+1} \\ a_{i+1}^TA^i & a_{i+1}^Ta_{i+1} \end{bmatrix}^{-1} , \tag{5.29}$$

which has the same form as (4.33) with G^i replacing B^i, $q = r = A^{iT}a_{i+1}$ and $\sigma = a_{i+1}^Ta_{i+1}$. Thus we can reformulate (4.25) in terms of G^{i+1} as

$$G^{i+1} = \overline{G}^i + ss^T , \tag{5.30}$$

where $\overline{G}^i$ is the matrix G^i bordered with an extra null row and column and

$$s = \delta \begin{bmatrix} G^iA^{iT}a_{i+1} \\ -1 \end{bmatrix} , \tag{5.31}$$

where δ is a scalar satisfying

$$\delta^2 = \frac{1}{a_{i+1}^Ta_{i+1} - a_{i+1}^TA^iG^iA^{iT}a_{i+1}} . \tag{5.32}$$

Note that Lemma 4.1 implies the positivity of the right-hand side in (5.32) and that the sign choice for δ is not important in the update.

The update of G^i using the above formula requires $i(i+1)/2$ multiplications in (5.30), having taken into account symmetry and $ni + i^2 + O(i)$ in (5.31). Thus the total number is $m^3/2 + nm^2/2$ plus lower-order terms.

From (5.28) we get for p_{i+1}

$$p_{i+1} = a_{i+1} - A^i G^i A^{iT} a_{i+1} \quad . \tag{5.33}$$

Since $u_1 = A^{iT} a_{i+1}$ and $u_2 = G^i u_1$ are needed for the update of G^i, the extra cost to be taken into account is only ni. Thus the total cost of this formulation is $m^3/2 + nm^2$. For $m = n$ this number is the same as in the standard formulation. For $m < n$ it is smaller, the standard formulation requiring $m(n + m)(n - m)/2$ more multiplications (say for $m = n/2$ there is a difference of a factor of three). For $m < n$ there is also a saving in memory occupation, since only $i(i + 1)/2$ positions are needed for G^i instead of the $n(n + 1)/2$ positions required by H_i.

If the modified Huang algorithm is considered, p_{i+1} is defined by

$$p_{i+1} = \bar{p}_{i+1} - A^i G^i A^{iT} \bar{p}_{i+1} \, , \tag{5.34}$$

where $\bar{p}_{i+1}$ is p_{i+1} defined by (5.33). Equation (5.34) requires $2in + i^2$ extra multiplications, thereby giving a total cost of $(5/6)m^3 + 2nm^2$. This number is smaller than in the standard formulation if $m < [(111^{1/2} - 6)/5]n \approx 0.9n$.

Let us consider the use of (4.38)–(4.40) (ALGORITHM 2). Since $H_i a_i = H_i^T a_i$, for the Huang algorithm we have the identity $s_i = p_i$; hence (4.38) becomes

$$p_i = a_i - \sum_{j=1}^{i-1} u_j p_j^T a_i. \tag{5.35}$$

Since in the Huang algorithm $w_i = a_i/a_i^T H_i a_i = a_i/a_i^T s_i = a_i/a_i^T p_i$, we have from (4.39)

$$u_i = \frac{p_i}{a_i^T p_i} \quad . \tag{5.36}$$

Using (5.36) in (5.35), we get

$$p_i = a_i - \sum_{j=1}^{i-1} \frac{p_j^T a_i}{p_j^T a_j} p_j \tag{5.37}$$

which is just another form of the unstabilized Gram–Schmidt iteration (5.4), since for the Huang algorithm $p_j^T a_j = a_j^T H_j a_j = a_j^T H_j^2 a_j = p_j^T p_j$. Huang (1975) gives arguments to suggest that (5.37) may be numerically preferable to (5.4).

For the modified Huang algorithm we have $z_i = H_i a_i = s_i$ and we can take $w_i = H_i a_i/(H_i a_i)^T (H_i a_i) = s_i/s_i^T s_i$. Thus we obtain the formulas

$$s_i = a_i - \sum_{j=1}^{i-1} s_j u_j^{\mathrm{T}} a_i \ , \tag{5.38}$$

$$p_i = s_i - \sum_{j=1}^{i-1} u_j s_j^{\mathrm{T}} s_i \ , \tag{5.39}$$

$$u_i = \frac{p_i}{s_i^{\mathrm{T}} s_i} \ . \tag{5.40}$$

Using (5.40) in (5.41) and (5.42), we obtain

$$s_i = a_i - \sum_{j=1}^{i-1} \frac{p_j^{\mathrm{T}} a_i}{s_j^{\mathrm{T}} s_j} s_j \ , \tag{5.41}$$

$$p_i = s_i - \sum_{j=1}^{i-1} \frac{s_j^{\mathrm{T}} s_i}{s_j^{\mathrm{T}} s_j} p_j \ . \tag{5.42}$$

In exact arithmetic, $s_i = p_i$; hence, if we substitute s_j, s_i by p_j, p_i, we obtain a formulation which needs to store only the vectors $p_1, \ldots, p_i$, given by

$$s_i = a_i - \sum_{j=1}^{i-1} \frac{p_j^{\mathrm{T}} a_i}{p_j^{\mathrm{T}} p_j} p_j \ , \tag{5.43}$$

$$p_i = s_i - \sum_{j=1}^{i-1} \frac{p_j^{\mathrm{T}} s_i}{p_j^{\mathrm{T}} p_j} p_j \ . \tag{5.44}$$

Note that (5.43) and (5.44) are equivalent to applying twice the unstabilized Gram–Schmidt procedure as suggested by Daniel *et al.* (1976). They are known as the reorthogonalized Gram–Schmidt procedure (Stoer and Bulirsch 1980).

Equations (5.41) and (5.42) require $2mn$ storage locations with a saving over the standard implementation if $m < n/2$ (the storage is halved if the Daniel *et al.* formulation is used). Since $4ni + \mathrm{O}(n)$ multiplications are required at the ith iteration, the total number of multiplications is $2nm^2 + \mathrm{O}(nm)$, less than in the standard implementation.

The Huang algorithm can be implemented in the form of ALGORITHM 3 by setting $u_j^1 = a_j$, $j = 1, \ldots, m$. The memory occupation is mn and the number of multiplications nm^2. It does not seem possible to write the modified Huang algorithm in the form of ALGORITHM 3, since the vectors $z_j = H_j a_j = u_j^1$ are not known in advance, save for $j = 1$.

Finally we can use (4.50) with $z_i = a_i$. Noting that $p_i^j = H_j z_i = H_j a_i$, we have the identity $a_j^T p_i^j = a_j^T H_j a_i = a_j^T H_j H_j z_i = p_j^T p_i^j$; also we have $a_j^T p_j = a_j^T H_j a_j = a_j^T H_j H_j p_j = p_j^T p_j$. Hence (4.50) can be written in the form

$$p_i^{j+1} = p_i^j - \frac{p_j^T p_i^j}{p_j^T p_j} p_i , \qquad j = 1, \ldots, i-1 , \qquad (5.45)$$

with $p_i^1 = a_i$. Note that (5.45) is identical with the stabilized Gram–Schmidt procedure, which is numerically preferable to the classical iteration (5.4), whose iterates p_i can quickly lose orthogonality (Rice 1966). The memory requirement is mn and the number of multiplications m^2n.

The given results on memory occupation and number of multiplications for different implementations are collected in Tables 5.1 and 5.2.

Table 5.1 — The Huang algorithm

Implementation	Memory	Multiplications
Standard	$n^2/2$	$(3/2)mn^2$
Equation (5.33)	$m^2/2$	$m^3/2 + m^2n$
ALGORITHM 2	mn	m^2n
ALGORITHM 3	mn	m^2n
Equation (5.45)	mn	m^2n

Table 5. — The modified Huang algorithm

Implementation	Memory	Multiplications
Standard	$n^2/2$	$(5/2)mn^2$
Equation (5.33)	$m^2/2$	$(5.6)m^3 + 2m^2n$
ALGORITHM 2	$2mn$	$2m^2n$
Equations (5.43) and (5.44)	mn	$2m^2n$

5.4   THE PARTIAL HUANG AND THE MODIFIED HUANG ALGORITHMS

We have seen in Chapter 3 (see Theorem 3.16 and Remark 3.8) that it is possible to choose z_{i+1} so that x_{i+2} satisfies not only the first $i+1$ equations but also the remaining equations so that termination occurs in $i+1$ steps. In order to obtain such an earlier termination, after having completed i iterations one has to compute the $m - i$ by n matrix $(H_{i+1}a_{i+1}, \ldots, H_{i+1}a_m)^{\mathrm{T}}$ and then solve for any solution s the full rank underdetermined system (3.52). Then setting $z_{i+1} = s$ terminates the algorithm at the iteration $i + 1$.

In this section we consider the application of this technique to force termination in the Huang and the modified Huang algorithms, aiming at algorithms with reduced computational cost or better numerical stability.

Definition 5.4
The i-partial Huang (modified Huang) algorithm is the variation of the Huang (modified Huang) algorithm where at the $(i+1)$th iteration the parameter z_{i+1} solves equation (3.52).

A natural question is whether there exists an optimal i-partial Huang (modified Huang) algorithm in the sense that the value of i minimizes the total number $\Phi(i, m, n)$ of multiplications required by the algorithm. Since Φ depends on how (3.52) is solved, the question can be considered only if a method for solving (3.52) is specified. Also we note that, for $i = 0$, (3.52) coincides with the original system, since $H_1 = I$, so that the question becomes in this case that of the optimal solver of a linear system.

It is natural to look for two solutions of (3.52): the basic type solution, where $n - m + i$ components of s are zero, and the minimal Euclidean norm solution, where $s = Z^{\mathrm{T}}d$, with d satisfying (3.49). The first solution can be obtained using Gaussian elimination (or, in the framework of the ABS methods, the implicit $LU - LL^{\mathrm{T}}$ method described in the next chapter) with column pivoting and requires $(m - i)^3/3 + O(mn)$ multiplications. The second solution can be obtained for instance by the Cholesky method on the normal-like equations (3.49), with a cost of $n(m - i)^2/2 + (m - i)^3/6 + O(mn)$ multiplications, or by the Huang algorithm applied on system (3.52) with the starting point s_1 proportional to the first row of Z.

If the Gauss or the Cholesky methods are used, it is clear that $i = 0$ is the optimal value for the i-partial algorithm, since these methods are less expensive on the original system. It is therefore of interest to consider whether there is any improvement in computational complexity by using the Huang or the modified Huang algorithms for solving (3.52). Since these algorithms can be implemented in different ways, say at least in the nine formulations considered in the previous section, there are at least 81 combinations that may be considered, depending on which algorithm is used on the original system up to the ith iteration and which one is used for solving (3.52). We limit our analysis to the following five combinations:

(a) The Huang standard–Huang standard.
(b) The modified Huang standard–modified Huang standard.
(c) The Huang ALGORITHM 2–Huang ALGORITHM 2.
(d) The modified Huang ALGORITHM 2–modified Huang ALGORITHM 2.
(e) The Huang ALGORITHM 3–Huang ALGORITHM 3.

In the following the number Φ is written as the sum of a term $\Phi_1(i, m, n)$ due to solving the first i equations of the original system and a term $\Phi_2(i, m, n)$ due to computing Z and solving (3.52). Only higher-order terms are considered. The optimal algorithm is determined by minimizing $\Phi(i, m, n)$ with respect to i.

In case (a) we have $\Phi_1 = (3/2)n^2 i$ and $\Phi_2 = (m - i)n^2 + (3/2)n^2(m - i)$. Thus $\delta\Phi/\delta i = -n^2$, implying that there is no benefit in forcing earlier termination, since the optimal value is $i = m$.

In case (b) we have $\Phi_1 = (5/2)n^2 i$ and $\Phi_2 = (m - i)n^2 + (5/2)n^2(m - i)$. Thus $\delta\Phi/\delta i = -n^2$ and we have the same case as before.

In case (c) we have $\Phi_1 = i^2 n$. Using (4.40) we have

$$H_{i+1}a_k = a_k - \sum_{j=1}^{i} s_j u_j^T a_k \ . \tag{5.46}$$

so that Z can be computed in $2in(m - i)$ multiplications. Thus $\Phi_2 = (m - i)^2 n + 2in(m - i)$ and $\delta\Phi/\delta i = 0$, again indicating that no benefit is produced by earlier termination.

A different result arises when we consider case (d). We have

$$\Phi_1 = 2i^2 n \ , \qquad \Phi_2 = 2n(m - i)^2 + 2ni(m - i) \text{ implying that}$$

$$\frac{\delta\Phi}{\delta i} = 2n(2i - m) \ . \tag{5.47}$$

Thus an optimal i-partial algorithm exists for $i = i^+$ given by

$$i^+ = \frac{m}{2} \ , \tag{5.48}$$

with a corresponding number of multiplications

$$\Phi(i^+, m, n) = \frac{3}{2}m^2 n + 0(mn) \ . \tag{5.49}$$

Thus a 25% reduction is obtained by forcing earlier termination.

Let us finally consider case (e). There is no need to compute Z, since the columns of Z coincide with the already available vectors u_j^{i+1}. In solving the first system, $2(m - j)n$ multiplications are required at the jth step, implying that $\Phi_1 = 2mni - ni^2$. Since $\Phi_2 = (m - i)^2 n$, we obtain $\delta\Phi/\delta i = 0$, so that the termination index has no effect on the computational cost.

An interesting observation can be made on the use of the Cholesky method for solving (3.52) via (3.49). While in general $i = 0$ is the optimal value (corresponding to a cost of $m^2 n/2 + m^3/6 + 0(mn)$ multiplications) in the case of the Huang method, in standard formulation a secondary optimal value exists if $m > (2^{1/2}-1)n$. Indeed we have $\Phi_1 = (3/2)in^2$, $\Phi_2 = (m - i)n^2 + n(m - i)^2/2 + (m - i)^3/6$ so that

$$\frac{\delta\Phi}{\delta i} = \frac{-i^2}{2} + i(m+n) + \frac{n^2}{2} - mn - \frac{m^2}{2} . \tag{5.50}$$

While for $i = m$ the derivative is positive, for $i = $ the derivative is negative if $m > (2^{1/2} - 1)n \approx 0.4n$. Thus a minimizer exists in the interval $[0, m]$ given by

$$i^+ = m - (2^{1/2} - 1)n . \tag{5.51}$$

The corresponding total number of multiplications is

$$\Phi(i^+, m, n) = \frac{3}{2}n^2m + \frac{n^3}{2}\left(-(2^{1/2}-1) + (2^{1/2}-1)^2 + \frac{(2^{1/2}-1)^3}{3}\right) . \tag{5.52}$$

Thus approximately $\Phi(i^+, m, n) = (3/2)n^2m - 0.1n^3$ or, for $m = n$, approximately $(7/5)n^3$; the greatest improvement is obtained for $m = (2^{1/2} - 1)n$, where the saving over the standard implementation is about 50%.

The following theorem shows that, when system (3.52) is solved by the Huang method, then the same search vectors are generated that would have been generated by the Huang method on the original system after the ith step. Thus the i-partial Huang algorithm with (3.52) solved via the Huang algorithm can be considered as another formulation of the basic Huang algorithm (with the benefit of a reduction in computational cost, at least in the case (d) considered above).

Theorem 5.7
Let $H_1, \ldots, H_{m+1}$ be the matrices generated by the Huang algorithm and let $\overline{H}_1, \ldots, \overline{H}_{1+m-i}$ be the matrices generated by the Huang algorithm on matrix Z in (3.48). Then for $j = 1, \ldots, m+1-i$ the identity $H_{i+j} = \overline{H}_j H_{i+1}$ is true.

Proof
We proceed by induction. For $j = 1$ the theorem is true, since $\overline{H}_1 = I$. Assume the $\overline{H}_j H_{i+1} = H_{i+j}$. Since $\overline{H}_j$ is updated using vector $\overline{H}_{i+1} a_{i+j}$ we have

$$\overline{H}_{j+1} H_{i+1} = \left(\overline{H}_j - \frac{\overline{H}_j H_{i+1} a_{i+j} a_{i+j}{}^T H_{i+1} \overline{H}_j}{a_{i+j}{}^T H_{i+1} H_j H_{i+1} a_{i+j})}\right) H_{i+1} . \tag{5.53}$$

Using the induction, we have

$$\overline{H}_{j+1} H_{i+1} = H_{i+j} - \frac{H_{i+j} a_{i+j} a_{i+j}{}^T H_{i+1} H_{i+j}}{a_{i+j}{}^T H_{i+1} H_{i+j} a_{i+j}} . \tag{5.54}$$

By Theorem 3.6, $H_{i+1} H_{i+j} = H_{i+j}$; hence $\overline{H}_{j+1} H_{i+1} = H_{i+j+1}$. Q.E.D.

Corollary 5.3
Let $p_1, \ldots, p_m$ be the search vectors generated by the Huang algorithm on matrix A and let $\bar{p}_1, \ldots, \bar{p}_{m-i}$ be the search vectors generated by the Huang algorithm on matrix Z in (3.48). Then for $j = 1, \ldots, m - i$ the identity $p_{i+j} = \bar{p}_j$ is true.

Remark 5.9
A consequence of Corollary 5.3 is that, instead of the special iteration with z_{i+1} being the solution of system (3.52), we can use the vectors $\bar{p}_j$ instead of the vectors p_{i+j} directly on the original system.

The above analysis suggests that we define the following ALGORITHM 4, here given without detailing the best memory arrangements.

ALGORITHM 4: The Fast Modified Huang Algorithm
(A4) Set $x_1 = \beta a_1$, with β an arbitrary scalar; set $s_1 = a_1$; set $i = 1$; set iflag $= 0$, $m^2 = [m/2]$; set for $j = 1, \ldots, m$, ivec$(j) = j$; define, for $j = 1, \ldots, m2$, the vectors $u_j = a_j$.

(B4) Compute the scalar $\tau_i = a_i^T x_i - b^T e_i$.

(C4) If $s_i \neq 0$ go to (D4); if $s_i = 0$ and $\tau_i = 0$ set $x_{i+1} = x_i$, iflag $=$ iflag $+ 1$ and stop if $i = m$ (x_{m+1} solves the system); otherwise set ivec$(i) = 0$ and go to (G4); if $s_i = 0$ and $\tau_i \neq 0$, set iflag $= -i$ and stop (system incompatible).

(D4) Compute the search vector p_i by

$$p_i = r_i - \sum_{j \in J(i-1)} \frac{s_j^T s_i}{s_j^T s_j} p_j \, , \tag{5.55}$$

where $J(i-1)$ is the set of nonzero values in ivec$(1), \ldots,$ ivec$(i-1)$.

(E4) Compute the new approximation x_{i+1} by

$$x_{i+1} = x_i - \frac{\tau_i}{p_i^T p_i} p_i \, . \tag{5.56}$$

If $i = m$, stop x_{u+1} solves the system.

(F4) Update the vector s_i by

$$s_{i+1} = u_i - \sum_{j \in J(i-1)} \frac{p_j^T u_i}{s_j^T s_j} s_j \, . \tag{5.57}$$

(G4) If i is not equal to $m2$, go to (H4); if $i = m2$ for $j = 1, \ldots, m - i$, define u_{m2+j} as follows:

$$u_{m2+j} = a_{m2+j} - \sum_{j \in J(i)} \frac{p_j^T a_{m2+j}}{s_j^T s_j} s_j . \tag{5.58}$$

Set $s_{i+1} = u_{m2+1}$ and, for $j = 1, \ldots, m2$ set $ivec(j) = 0$.

(H4) Increment the index i by one and go to (B4).

5.5 RELATIONSHIPS WITH THE MOORE–PENROSE PSEUDOINVERSE

The subsystem consisting of the first i equations of system $Ax = b$ has the form

$$A^{iT}x = b^i, \tag{5.59}$$

where $b^i \in R^i$ is the vector whose components are the first i components of b. The minimal Euclidean norm solution of (5.59) can be expressed in terms of the Moore–Penrose pseudoinverse, (see for instance Albert (1972)), by

$$\bar{x} = (A^{iT})^+ b^i . \tag{5.60}$$

The general solution of (5.59) can also be expressed in terms of x and $(A^{iT})^+$ by

$$x = \bar{x} + [I - (A^{iT})^+ A^{iT}] y \tag{5.61}$$

with $y \in R^n$ an arbitrary vector.

If A^i is full rank, then $(A^{iT})^+$ is given by the well-known formula

$$(A^{iT})^+ = A^i (A^{iT} A^i)^{-1} . \tag{5.62}$$

Thus, if H_{i+1} is the matrix obtained at the ith step of the Huang algorithm applied to system (5.59), we obtain from (5.27) and (5.62) the identity

$$H_{i+1} = I - (A^{iT})^+ A^{iT} . \tag{5.63}$$

The Moore–Penrose pseudoinverse can be easily computed in the framework of the Huang algorithm. Indeed, for $j = 1, \ldots, i$, let b_j^i be i linearly independent vectors and let $\bar{x}^j$ be the solution of the system $A^{iT}x = b_j^i$, computed by the Huang algorithm with x_1 proportional to a_1. Since, by Theorem 5.4, $\bar{x}^j$ is the minimal Euclidean norm solution of $A^{iT}x = b_j^i$, we obtain from (5.60) the identity

$$\bar{x}^j = (A^{iT})^+ b_j^i . \tag{5.64}$$

Let X be the n by i matrix whose jth column is the vector $\bar{x}^j$ and let B be the i by i matrix whose jth column is b_j^i. Then (5.64) implies that

$$(A^{iT})^+ = XB^{-1} \tag{5.65}$$

or

$$(A^i)^+ = (B^{-1})^T X^T \ . \tag{5.66}$$

If we take $b_j^i = e_j^i$ as the jth unit vector in R^i, then the above formulas simplify; $\bar{x}_j$ is then the jth column of $(A^{iT})^+$. Note that, since the same search vectors are used to solve the i systems, once the first system is solved, the only computations required for the remaining systems are those for updating the iterates x_k. If $b_j^i = e_j^i$ and the systems are solved in the order $j = i, i-1, \ldots, 1$, then the first j iterates x_k of the last $i-j+1$ systems are identical. If the first j iterates x_k are stored, then, in solving the jth system, only the last $i-j+1$ iterates x_k have to be computed, with a cost of $2n(i-j) + 0(n)$ multiplications. Thus the Moore–Penrose pseudoinverse can be evaluated with an additional cost of no more than $ni^2 + 0(in)$ multiplications over that required for solving a system of i equations in n unknowns.

5.6 RELATIONSHIPS WITH THE ALGORITHM OF PYLE AND THE ALGORITHM OF BRENT

The Huang algorithm is equivalent, in the sense of generating the same set of iterates x_i, to two algorithms which predate the Huang (1975) paper: the algorithm of Pyle, proposed in 1964, and the algorithm of Brent, published in 1973. The Pyle algorithm has been given in the framework of the gradient projection method for linearly constrained optimization to compute the minimal Euclidean norm solution of underdetermined systems; it has been used by Albert (1972) for the computation of the pseudoinverse. The Brent algorithm has been proposed for the solution of nonlinear systems as an alternative method, preferable in terms of numerical stability, to the Brown (1969) method. It can be considered as a method for linear equations, not only because these are a special case of nonlinear equations but also because it has finite termination in such a case.

The Pyle method for solving the system $Ax = b$, assumed to be compatible, consists of the following steps.

The Pyle Algorithm

(A) Set $x_1 = 0$; set $i = 1$; let $d_1, \ldots, d_m$ be the vectors obtained by applying the Gram–Schmidt orthogonalization procedure (5.4) to the rows of A (omitting the kth term in the summation, if $d_k = 0$ owing to the linear dependence of the kth equation).

(B) Update the estimate of the solution by

$$x_{i+1} = x_i - \alpha_i d_i \ , \tag{5.67}$$

with $\alpha_i = 0$ if $d_i = 0$; otherwise $\alpha_i = (a_i^T x_i - b^T e_i)/a_i^T d_i \ .$

(C) If $i = m$, stop (x_{m+1} equals A^+b); otherwise increment the index i by one and go
to (B).

It is clear that the Pyle algorithm is equivalent to the Huang algorithm, since both
methods use the same search vectors at the ith step. For other approaches to
computing the minimal Euclidean norm solution of underdetermined systems see
Cline and Plemmons (1976). The procedure proposed by Albert (1972) to compute
the pseudoinverse in the framework of the Pyle method is the following. Let
$d_1, \ldots, d_q$, $q = \text{rank}(A)$, be the set of nonzero normalized vectors obtained by
applying the Gram–Schmidt procedure to the columns of A. Then by a property of
the Moore–Penrose pseudoinverse the following identity is true

$$AA^+ = \sum_{j=1}^{q} d_j d_j^{\mathrm{T}} . \tag{5.68}$$

Let $b_1, \ldots, b_m$ be the columns of AA^+ defined as above. Since $\text{Range}(AA^+) =$
$\text{Range}(A)$, the systems $Ax = b_j$ are compatible for $j = 1, \ldots, m$. Let $\bar{x}_j = A^+b_j$ be the
solution of these systems computed by the Pyle method. Then, since
$(A^+b_1, \ldots, A^+b_m) = A^+(b_1, \ldots, b_m) = A^+(AA^+) = A^+$, the vector x_j is the jth
column of A^+. We can observe that the procedure proposed by us in the previous
section is less expensive and probably more stable since it requires orthogonalization
of only the rows of A and not also of the columns.

The Brent (1973) algorithm for solving the nonlinear system $f(y) = 0$, f, $y \in R^n$,
generates a sequence of iterates y_k defined in the following way: y_1 is given and y_{k+1}
is obtained through the following steps, which define an inner cycle:

The Brent Algorithm
(A) Set $\bar{x}_1 = y_k$; set $i = 1$.
(B) Define the vector $a_i \in R^n$ as the ith row of the Jacobian of f in $\bar{x}_i$, i.e. for
$j = 1, \ldots, n$

$$a_i^{\mathrm{T}} e_j = \frac{\delta f_i(\bar{x}_i)}{\delta x_j} . \tag{5.69}$$

(C) Let $\bar{x}_{i+1}$ be the solution of the linear system

$$a_j^{\mathrm{T}} x = a_j^{\mathrm{T}} \bar{x}_j - e_j^{\mathrm{T}} f(\bar{x}_j) , \qquad j = 1, \ldots, i , \tag{5.70}$$

which is closest to $\bar{x}_i$ in the Euclidean norm.
(D) If $i = n$, set $y_{k+1} = \bar{x}_{n+1}$ and stop; otherwise increment the index i by one and go
to (B).

The inner cycle in the Brent method corresponds to solving a linear system $Ax =$
b whose equations are dynamically defined as the algorithm proceeds, since at the ith

step the ith row of A and the ith component of b are determined in terms of the computed solution of the first $i-1$ equations.

It is clear that the Huang method can be applied to solve such a linear system, since at the ith step it uses only the ith equation, which is defined at the beginning of that step. Moreover, if the initial vector x_1 in Huang method is equal to y_k, then for $i=1,\ldots,n$, $x_{i+1} = \bar{x}_{i+1}$ since in the Huang algorithm, by Theorem 5.3, x_{i+1} is the unique vector closest in euclidean norm to x_i, as is true for the corresponding vector in the Brent cycle.

The procedure given by Brent and refined by Moré and Cosnard (1979) to compute $\bar{x}_{i+1}$ in step (C) is based upon the explicit construction of certain orthogonal matrices. At the beginning of the ith step an n by n orthogonal matrix Q_{i+1} is given such that the following relation is satisfied:

$$A^{iT}Q_{i+1} = [L_i, 0] , \tag{5.71}$$

with L_i an i by i nonsingular lower triangular matrix. The ith column of Q_{i+1} is taken as a search vector p_i. Then $\bar{x}_{i+1} = \bar{x}_i - \alpha_i p_i$, with the step size α_i given by

$$\alpha_i = \frac{e_i^T f(\bar{x}_i)}{\sigma_i} , \tag{5.72}$$

with σ_i the ith diagonal element of L_i, i.e. $\sigma_i = a_i^T p_i$ Q_1 can be taken as an arbitrary orthogonal matrix (but the cases $Q_1 = I$ and Q_1 equal to the final matrix of the previous cycle are of special interest) and Q_{i+1} is built up from

$$Q_{i+1} = Q_i U_i , \tag{5.73}$$

where U_i is an orthogonal matrix of the form

$$U_i = \begin{bmatrix} I_{i-1} & 0 \\ 0 & U \end{bmatrix} , \tag{5.74}$$

such that the vector $U_i^T Q_i^T a_i$ has the ith component equal to $\sigma_i \neq 0$ and null components from the index $i+1$ to n. The matrix U_i can be defined either as a Householder matrix or as the product of $n-i$ Givens rotations. In the first case the update of Q_i requires $n^3 + 0(n^2)$ multiplications and n square roots; the storage requirement is $n^2 + 0(n)$ positions. In the second case $n^3 + 0(n)$ multiplications and $n^2/2 + 0(n)$ square roots are needed; if $Q_1 = I$ the storage requirement can be reduced to $n^2/2 + 0(n)$.

We note that since on linear systems $e_i^T f(\bar{x}_i) = a_i^T \bar{x}_i - b^T e_i$ the step size formula (5.72) coincides with the step size formula of the ABS class. If the Huang method is used to solve the linear system in Brent inner cycle, then the search vector defined in the ith iteration is parallel to the ith column of Q_{i+1}, but generally is different in norm. It is clear, however, that the matrix Q_{i+1} in the Brent cycle is the same as the matrix Q_i defined in (5.1) for the Huang algorithm. In a sense in the Brent inner cycle an approximation is available at the ith step of future search vectors, through the $(i + 1)$th to the nth columns of Q_{i+1}.

5.7 BIBLIOGRAPHICAL REMARKS

In his seminal paper, Huang (1975) proved a number of results here given in Corollary 5.1, Remark 5.2 and Theorems 5.2 and 5.5. Theorem 5.4 generalizes Huang's result, which was given only for the case $x_1 = 0$. Theorem 5.1 also generalizes Huang's result to the case $H_1 \neq I$; it was first given by Abaffy and Spedicato (1982). Theorems 5.3 and 5.6 are due to Spedicato and are stated here for the first time. Corollary 5.2 has appeared in the paper by Abaffy and Spedicato (1985). The modified Huang algorithm is due to Spedicato and has been studied numerically in various papers by Abaffy and Spedicato (1983a, 1987), Bertocchi and Spedicato (1988a), Spedicato and Vespucci (1989) and Spedicato and Bodon (1989b). The partial Huang and the modified Huang algorithm are due to Spedicato and are presented here for the first time. The literature on the pseudoinverse is immense (see for instance Albert (1972), Bouillon and Odell (1971), Rao and Mitra (1971), Nashed (1976) and Zielke (1984)).

6

The implicit Gauss–Cholesky (implicit LU–LL^T) algorithm

6.1 INTRODUCTION

In this chapter we consider another fundamental algorithm in the ABS class, namely the algorithm that generates an implicit factorization of the matrix A in the LU form (when A is symmetric positive definite in the LL^T form). Clearly such an algorithm can be interpreted as an alternative formulation of the classical LU or Gaussian elimination algorithm (when A is symmetric positive definite, of the Cholesky algorithm). A remarkable property of the algorithm is that, to obtain the implicit LL^T factorization when A is symmetric positive definite, no change is needed in the implicit LU algorithm. Thus the name implicit LU–LL^T algorithm appears to be appropriate.

As was the case for the Huang algorithm, the implicit LU–LL^T algorithm can be implemented in an infinite number of ways. We shall be mainly concerned with an implementation that has the same order of cost as the classical LU algorithm, say $n^3/3 + \mathrm{O}(n^2)$ multiplications. Intriguingly it seems that, when A is symmetric positive definite, this cost cannot be reduced, as it can in the classical Cholesky algorithm, to $n^3/6 + \mathrm{O}(n^2)$. However in such a case the additional cost of inverting a matrix is $n^3/6 + \mathrm{O}(n^2)$, against the number $n^3/3 + \mathrm{O}(n^2)$ required classically. Thus the order of cost of inverting a positive definite matrix appears to be an invariant quantity in both the classical and the ABS approach.

The implicit LU–LL^T algorithm is related to other algorithms appearing in the literature. It is equivalent, when the starting point is the null vector, to the classical escalator method, in the sense of generating the same set of iterates x_i. In the same sense it is also equivalent to the Brown (1969) method for nonlinear systems, when applied to a linear system, and to a special method in a general class proposed by Sloboda (1978) for solving linear systems.

6.2 DERIVATION OF THE PARAMETERS THAT GENERATE AN IMPLICIT *LU* FACTORIZATION

It is well known that a nonsingular square matrix A can be factorized in the form $A = LU$ with L and U respectively lower and upper triangular nonsingular matrices, if and only if it is strongly nonsingular. If A is nonsingular but not strongly nonsingular, permutation matrices P_1, P_2 exist such that $P_1 A$ or AP_2 is strongly nonsingular; thus a nonsingular matrix A can always be factorized in the form $A = LU$ after suitable permutation of the rows or the columns.

A natural question in the context of the ABS class is the following: given A, a square nonsingular matrix, is it possible to choose the parameters H_1, z_i, w_i so that in the underlying factorization (3.39) the matrix $S = P^{-1}$ is upper triangular? The answer, a qualified yes, is given in the form of a sufficient condition by the following.

Theorem 6.1
Let A be square nonsingular and assume that the matrix H_1 and the vectors $w_1, \ldots, w_n$ are such that:

(a) condition (3.6) is satisfied and
(b) the matrix $H_1 W^n$ is upper triangular, where $W^n = (w_1, \ldots, w_n)$.

Then the choice $z_i = w_i$ satisfies condition (3.2) and with this choice the matrix $P^n = (p_1, \ldots, p_n)$ is upper triangular.

Proof
From assumption (a) it follows immediately that (3.2) is satisfied. For the given choice of z_i we have $p_i = H_i^T w_i$ and therefore $P^i = \overline{W}^i$ with $\overline{W}^i$ defined as in (3.33). For $i = n$ from the proof of Theorem 3.10 we have

$$(P^n)^T A = R_1, \tag{6.1}$$

with R_1 upper triangular, and

$$(W^n)^T \overline{A}^n = L_1, \tag{6.2}$$

with $L_1 = R_2^T$ lower triangular. From (6.1) and (6.2) we obtain

$$(W^n)^T \overline{A}^n (P^n)^T A = L_1 R_1 . \tag{6.3}$$

From Theorem 4.1 we have that $\overline{A}^n (P^n)^T = \overline{A}^n (\overline{W}^n)^T = H_1$, since H_{n+1} is the null matrix. Thus (6.3) gives

$$(W^n)^T H_1 A = L_1 R_1 . \tag{6.4}$$

Expressing A from (6.4) and substituting in (6.1) give

$$(P^n)^T = L_1^{-1} (W^n)^T H_1 . \tag{6.5}$$

In the underlying factorization (3.39) the matrix $S = P^{-1} = (P^n)^{-1}$ is upper triangular if and only if $(P^n)^T$ is lower triangular. From (6.5) a sufficient condition for this to be true is that $(W^n)^T H_1$ be lower triangular, which is true by assumption. Q.E.D.

Remark 6.1
Conditions (a) and (b) of Theorem 6.1 can be satisfied only if A is strongly nonsingular, since otherwise Theorem 6.1 would provide a construction of an LU factorization of A, which is impossible from classical results.

   The following theorem shows, as expected, that a row pivoting strategy always exists such that assumptions (a) and (b) of Theorem 6.1 are satisfied by some choices of H_1 and W^n.

Theorem 6.2
Let A be square nonsingular and let $H_1{}^T$ and $\overline{W}^n = (\overline{w}_1, \ldots, \overline{w}_n)$ be nonsingular upper triangular matrices. Then for $i = 1, \ldots, n$ it is possible to choose an index j with $i \leqslant j \leqslant n$ and a nonzero scalar β_i such that $w_i{}^T H \overline{a}_j = 1$, with $w_i = \beta_i \overline{w}_i$ and $\overline{a}_j$ the jth row of a matrix A obtained after some interchanges of the rows of A.

Proof
We proceed by induction and contradiction. For $i = 1$ we have $w_1 = \tau_1 e_1$ and $H_1{}^T e_1 = \tau_2 e_1$ with τ_1, τ_2 nonzero scalars. Thus $w_1{}^T H_1 a_j = \tau_1 \tau_2 e_1^T a_j$ must be nonzero for at least one index j, since not all elements of the first column of A can be zero. Let $\bar{j}$ be the first index j for which the above scalar product is nonzero and let $\overline{A}_1$ be the matrix obtained by interchanging the first and the $\bar{j}$th row of A; set $w_1 = \overline{w}_1 / \overline{w}_1{}^T H_1 \overline{a}_j$. Then $w_1{}^T H_1 \overline{a}_1 = 1$ and the theorem is true for $i = 1$. Assume now that the theorem is true up to the index i. Let $\overline{A}_i = (\overline{A}^i, \overline{A}^{n-i})$ be the matrix obtained at the ith step from A after the required row permutations have been performed, $\overline{A}^i$ comprising the first i columns and $\overline{A}^{n-i}$ the last $n - i$ columns. Let H_{i+1} be the matrix available at the end of the ith step and observe that H_{i+1} satisfies relation $H_{i+1} \overline{A}^i = 0$. We complete the induction by showing that a contradiction arises if the theorem is not true for the index $i + 1$. If at the $(i+1)$th step it is impossible to choose an index j, $i + 1 \leqslant j \leqslant n$, such that $w_{i+1}{}^T H_{i+1} a_j \neq 0$, we would have

$$\overline{w}_{i+1}{}^T H_{i+1} \overline{A}^{n-i} = 0 \tag{6.6}$$

or, since $H_{i+1} \overline{A}^i = 0$,

$$\overline{w}_{i+1}{}^T H_{i+1} A^n = 0 . \tag{6.7}$$

Since nonsingularity of A implies nonsingularity of $\overline{A}^n$, (6.7) implies that $\overline{w}_{i+1}$ lies in the null space of $H_{i+1}{}^T$. Since such a space is spanned by $w_1, \ldots, w_i$, w_{i+1} must be of the form

$$\overline{w}_{i+1} = W^i q \tag{6.8}$$

for some $q \in R^i$. Since W^i is formed by the first i columns of $\overline{W}^n$ suitably scaled, only the first i components of $W^i q$ can be nonzero. By assumption, $\overline{w}_{i+1}$ is the $(i+1)$th column of a nonsingular triangular matrix. Thus its $(i+1)$th component must be nonzero, evidencing a contradiction. It follows that a first index $\overline{j}, i+1 \leqslant \overline{j} \leqslant n$, exists such that $\overline{w}_{i+1}{}^T H_{i+1} \overline{a}_{\overline{j}} \neq 0$. Setting $w_{i+1} = \beta_{i+1} \overline{w}_{i+1}$ with $\beta_{i+1} = 1/\overline{w}_{i+1}{}^T H_{i+1} \overline{a}_{\overline{j}}$ and $\overline{A}_{i+1}$ equal to $\overline{A}_i$, apart from the interchange of the $(i+1)$th column with the $\overline{j}$th one, completes the induction. Q.E.D.

Theorem 6.2 has shown the existence of a whole class of implicit LU factorization algorithms, where H_1 and W^n are upper triangular nonsingular matrices (with possible row pivoting for well definiteness). The given choices for H_1 and W^n induce a structure in the matrix H_i, described by the following theorem.

Theorem 6.3
Let $H_1{}^T$ and W^n be nonsingular upper triangular matrices and consider the sequence of matrices generated by (3.5) with A strongly nonsingular and $w_i = e_i/e_i{}^T H_i a_i$. Then the following properties are true.

(a) The first i rows of H_{i+1} are identically zero.
(b) The last $n - i$ columns of H_{i+1} are equal to the last $n - i$ columns of H_1.

Proof
By Theorem 3.8 we have $H_{i+1}{}^T w_j = 0$ for $j = 1, \ldots, i$. For $j = 1$ we have $w_1 = H_{1,1} e_1$, $H_{1,1} \neq 0$, implying that $H_{i+1}{}^T e_1 = 0$, i.e. the first row of H_{i+1} is zero. Since $w_j = \tau_{1,j} e_1 + \ldots + \tau_{j,j} e_j$ with $\tau_{j,j} \neq 0$, induction on j establishes that $H_{i+1}{}^T e_j = 0$ for $j \leqslant i$, proving (a). Now $H_i{}^T w_i = \tau_{i,i} H_i{}^T e_i$, $\tau_{i,i} \neq 0$, so that the update for H_i reads

$$H_{i+1} = H_i - \frac{H_i a_i e_i{}^T H_i}{e_i{}^T H_i a_i} . \tag{6.9}$$

Since the first $i - 1$ rows of H_i are zero by property (a) it follows that the update (6.9) changes only those elements of H_i that belong to the intersection of the first i columns with the last $n - i + 1$ rows (with the effect of zeroing the ith row), proving statement (b). Q.E.D.

Remark 6.2
From Theorem 6.3 we can write H_{i+1} as follows

$$H_{i+1} = \begin{bmatrix} 0 & 0 \\ S_i & (H_1)_i \end{bmatrix}, \tag{6.10}$$

where S_i has $n - i$ rows and i columns and $(H_1)_i$ is the submatrix of H_1 consisting of the intersection of the last $n - i$ rows and columns.
Since $H_i{}^T w_i = \tau_{i,i} H_i{}^T e_i$ no loss of generality occurs if we set $W^n = I$ and $z_i = e_i$.

This leads to the following formulas for the computation of the search vector and the update of the matrix (for A strongly nonsingular):

$$p_i = H_i^T e_i \ , \tag{6.11}$$

$$H_{i+1} = H_i - \frac{H_i a_i p_i^T}{p_i^T a_i} \ . \tag{6.12}$$

Remark 6.3
Since H_1^T is arbitrary nonsingular upper triangular (6.11) and (6.12) define a class of algorithms generating an implicit LU factorization.

Definition 6.1
The subclass of the ABS class defined by (6.11) and (6.12) with H_1^T nonsingular upper triangular and A strongly nonsingular is called the subclass of the implicit LU factorization or the implicit Gaussian elimination.

Definition 6.2
The algorithm in the subclass of the implicit LU factorization where $H_1 = I$ is called the implicit LU factorization algorithm.

Theorem 6.4
If no pivoting is required, the implicit LU factorization algorithm can be implemented with no more than $n^3/3 + O(n^2)$ multiplications.

Proof
First observe that no computations are required for evaluating p_i by (6.11) and that p_i has at most i nonzero components. In the evaluation of H_{i+1} no more than $(n-i)(i+1)$ multiplications are required for computing $H_i a_i$, since $(H_1)_i$ in (6.10) is the unit matrix in R^{n-i}; no more than i multiplications and i divisions are required for computing $s_i = p_i/p_i^T a_i$; no more than $(n-i)i$ multiplications are required for computing the nonzero elements of $H_i a_i s_i^T$, say the matrix S_i in (6.10). The theorem follows by summing all terms. Q.E.D.

Theorem 6.5 establishes that, as expected, no pivoting is required if the matrix A is strongly nonsingular. In the proof we shall use the following

Lemma 6.1
Let A be a strongly nonsingular matrix. Let a_{i+1} and $\bar{a}_{i+1}$ be respectively the $(i+1)$th row and column. Define the matrix K_i by

$$K_i = \begin{bmatrix} (A^{i,i})^{-1} & 0 \\ 0 & 0 \end{bmatrix} , \tag{6.13}$$

where $A^{i,i}$ is the ith principal submatrix of A. Then

$$e_{i+1}{}^{\mathrm{T}}a_{i+1} - a_{i+1}{}^{\mathrm{T}}K_i\bar{a}_{i+1} \# 0 \tag{6.14}$$

Moreover if A is symmetric positive definite,

$$e_{i+1}{}^{\mathrm{T}}a_{i+1} > a_{i+1}{}^{\mathrm{T}}K_i\bar{a}_{i+1} \tag{6.15}$$

Proof
It is enough to apply Lemma 4.1 with the definition $\sigma = e_{i+1}{}^{\mathrm{T}}a_{i+1}$ and to note that, in the product $a_{i+1}{}^{\mathrm{T}}K_i\bar{a}_{i+1}$ only the first i components of $a_{i+1}, \bar{a}_{i+1}$ are used. Q.E.D.

Theorem 6.5
Let A be a strongly nonsingular matrix. Then the sequence of matrices generated by (3.5), with H_1 the unit matrix and $w_i = e_i/\delta_i$, $\delta_i = e_i^{\mathrm{T}}H_i a_i$, is well defined.

Proof
It is enough to show that all denominators δ_i are nonzero and we prove this by induction. For $i = 1$, δ_i is nonzero, since $e_1{}^{\mathrm{T}}a_1 = A^{1,1}$ is nonzero. Assume now that the sequence is well defined up to the index i. We note that the given choices of H_1 and w_j, $j = 1, \ldots, i$, imply the relation

$$(W^i)^{\mathrm{T}}H_1 A^i = D_i(A^{i,i})^{\mathrm{T}}, \tag{6.16}$$

where D_i is the nonsingular (by induction) diagonal matrix of order i whose jth diagonal element is δ_j, and $A^{i,i}$ is the nonsingular (by assumption) ith principal submatrix of A. Using (4.6) and (6.16), we can write H_{i+1} as follows:

$$H_{i+1} = I - A^i(A^{i,i})^{-\mathrm{T}}E_i , \tag{6.17}$$

where E_i is the i by n matrix whose jth row is the jth unit vector in R^n. Let $\bar{a}_{i+1}$ be the $(i+1)$th column of A, let $\bar{a}_{i+1}' \in R^i$ be the vector consisting of the first i elements of $\bar{a}_{i+1}$; let $a_{i+1}' \in R^i$ be the vector consisting of the first i elements of a_{i+1}. Observe that we have the identities

$$E_i a_{i+1} = a_{i+1}' \tag{6.18}$$
$$A^{i\mathrm{T}}e_{i+1} = \bar{a}_{i+1}'. \tag{6.19}$$

Multiplying (6.17) on the left by e_{i+1} and on the right by a_{i+1}, we have, using (6.18) and (6.19),

$$\delta_{i+1} = e_{i+1}{}^{\mathrm{T}}a_{i+1} - \bar{a}_{i+1}'^{\mathrm{T}}(A^{i,i})^{-\mathrm{T}}a_{i+1}' . \tag{6.20}$$

Observing that $\bar{a}_{i+1}'^{\mathrm{T}}(A^{i,i})^{-\mathrm{T}}a_{i+1}' = a_{i+1}'^{\mathrm{T}}(A^{i,i})^{-1}a_{i+1}'$ and that, with K_i defined as in Lemma 6.1, we get

$$K_i a_{i+1} = \begin{bmatrix} (A^{i,i})^{-1} a_{i+1}' \\ 0 \end{bmatrix} ,$$

(6.21)

we can write (6.20) as follows:

$$\delta_{i+1} = e_{i+1}{}^T a_{i+1} - a_{i+1}{}^T K_i \bar{a}_{i+1}$$

(6.22)

and the thesis follows from Lemma 6.1. Q.E.D.

Definition 6.3
The scalar $\delta_i = e_i{}^T H_i a_i$ defined in the implicit LU factorization algorithm is called the ABS pivot of the implicit LU factorization.

Corollary 6.1
If A is strongly nonsingular, then the ABS pivots are all nonzero; if moreover A is symmetric positive definite, then the ABS pivots are all positive.

Corollary 6.2
The necessary and sufficient condition for the implicit LU factorization algorithm to be well defined without row pivoting is that A be strongly nonsingular.

Theorem 6.6
Let A be strongly nonsingular and let $H_1 = I$. Then the ABS pivots satisfy relations $\delta_1 = \det(A^{1,1})$ and for $i > 1$

$$\delta_i = \frac{\det(A^{i,i})}{\det(A^{i-1,i-1})} .$$

(6.23)

Proof
For $i = 1$ the theorem is trivial and for $i > 1$ it follows from (4.22). Q.E.D.

Corollary 6.3
The determinant of a strongly nonsingular matrix is equal to the product of the ABS pivots . The modulus of the determinant of a nonsingular matrix is equal to the modulus of the product of the ABS pivots.

6.3 PROPERTIES OF THE VECTORS p_i AND x_i

In this section, we establish some remarkable properties of the search vectors p_i and of the approximations x_i to the solution. The main property of the search vectors is that they are A-semiconjugate (or A-conjugate if A is symmetric). Thus the implicit LU algorithm can be viewed as a procedure for building sets of A-semiconjugate (or

A-conjugate) vectors in $n^3/3 + O(n^2)$ multiplications (if no pivoting is required). Related to this property is the variational characterization of the vectors x_i in terms of minimization of a quadratic function on spaces of increasing dimensions.

Theorem 6.7

Let A be strongly nonsingular and let $p_1, \ldots, p_n$ be the search vectors generated by the implicit LU algorithm. Then the following A-semiconjugacy relation is true:

$$p_i^T A p_j = 0 \qquad 1 \leqslant i \leqslant j - 1, \quad j = 2, \ldots, n \ , \tag{6.24}$$

$$p_i^T A p_i \neq 0, \qquad i = 1, \ldots, n \ . \tag{6.25}$$

Proof

Observe that $p_i^T A p_j = e_i^T H_i (A^n)^T H_j^T e_j = e_i^T H_i (H_j A^n)^T e_j$. Since $A^n = (A^{j-1}, A^{n-j+1})$ and $H_j A^{j-1}$ is zero, from (3.27), we obtain

$$p_i^T A p_j = e_i^T H_i (0, H_j A^{n-j+1})^T e_j$$

or

$$p_i^T A p_j = u_i^T q_j \ , \tag{6.26}$$

where $q_j = (0, H_j A^{n-j+1})^T e_j$ is a vector of the form

$$q_j = \begin{bmatrix} 0 \\ \cdot \\ 0 \\ e_j^T H_j a_j \\ \cdot \\ e_j^T H_j a_n \end{bmatrix} \tag{6.27}$$

and $u_i = H_i^T e_i$ has the following form, from the structure of H_i, (see (6.10)):

$$u_i = \begin{bmatrix} e_1^T (S_{i-1})^T e_i \\ \cdot \\ e_{i-1}^T (S_{i-1})^T e_i \\ 1 \\ 0 \\ \\ 0 \end{bmatrix} \tag{6.28}$$

Since $i \leqslant j - 1$, the scalar product $u_i^T q_j$ is identically zero. Since $p_i^T A p_i = u_i^T q_i = e_i^T H_i a_i$ is nonzero from Corollary 6.1 the theorem follows. Q.E.D.

Corollary 6.4
If A is symmetric strongly nonsingular, then the search vectors generated by the implicit LU algorithm are A-conjugate.

Remark 6.4
Note that positive definiteness is redundant for the generation of A-conjugate vectors via the implicit LU algorithm.

We give now some theorems characterizing the sequence of the approximations x_i to the solution.

Theorem 6.8
Let x_i, x_{i+1} be successive vectors generated by the implicit LU factorization algorithm. Then the last $n - i$ components of the vector $x_{i+1} - x_i$ are zero.

Proof
The vector $x_{i+1} - x_i$ is proportional to the ith row of H_i and the theorem follows from the structure of H_i, (see (6.10)). Q.E.D.

Remark 6.5
Theorem 6.8 implies that, if x_1 is the null vector, the solution is approximated through a sequence of solutions x_i of the first $i - 1$ equations of the basic type (whose nonzero components lie in the first $i - 1$ positions).

The following theorem gives a variational characterization of the step size α_i and of the approximations x_i of the solution.

Theorem 6.9
Let A be symmetric positive definite. Then the step size α_i in the implicit LU factorization algorithm minimizes along the line $x(\alpha) = x_i - \alpha p_i$, the squared A-weighted Euclidean distance from the solution $x^+ = A^{-1}b$, i.e. the convex quadratic function

$$F(\alpha) = [x(\alpha) - x^+]^T A [x(\alpha) - x^+] \ . \tag{6.29}$$

Proof
Along the line $x(\alpha)$ the derivative of F with respect to α is given by

$$\frac{\delta F(\alpha)}{\delta \alpha} = -p_i^T \operatorname{grad}\{F[x(\alpha)]\} \ . \tag{6.30}$$

Since $\quad \text{grad}\{F[x(\alpha)]\} = 2A[x(\alpha) - x^+] = 2Ax(\alpha) - 2b = 2(Ax_i - b - \alpha Ap_i) = 2(r_i - \alpha Ap_i)$ with r_i the residual vector in x_i, we obtain

$$\frac{\delta F(\alpha)}{\delta \alpha} = -2p_i^T(r_i - Ap_i) \ . \tag{6.31}$$

Setting the derivative to zero gives the following value for the minimizer α^+:

$$\alpha^+ = \frac{p_i^T r_i}{p_i^T Ap_i} \ . \tag{6.32}$$

It is now enough to prove the following identities:

$$p_i^T r_i = a_i^T x_i - b^T e_i \ , \tag{6.33}$$

$$p_i^T Ap_i = a_i^T p_i \ . \tag{6.34}$$

Equation (6.33) follows from the observation that the first $i - 1$ components of r_i are zero (for any algorithm of the ABS class) and, from (6.10), that the last $n - i$ components of p_i are zero, while the ith component is equal to one. To establish (6.34), we note that $Ap_i = AH_i^T e_i = (H_i A^n)^T e_i = (H_i A^{i-1}, H_i a_i, H_i A^{n-i})^T e_i$. From Theorem 3.3, $H_i A^{i-1}$ is zero, hence Ap_i has the form

$$Ap_i = \begin{bmatrix} 0 \\ \dots \\ 0 \\ a_i^T p_i \\ (A^{n-i})^T H_i e_i \end{bmatrix} \tag{6.35}$$

and (6.34) follows from the structure of p_i. Q.E.D.

Theorem 6.10
Let A be symmetric positive definite. Then the vector x_{i+1} obtained by the implicit LU factorization algorithm minimizes the weighted Euclidean norm (6.29) of the distance from the solution over the linear variety $x_1 + \text{Span}(p_1, \dots, p_i)$.

Proof
From Theorem 6.7 the vectors $p_1, \dots, p_i$ are A-conjugate; from Theorem 6.9 the vector x_{i+1} is obtained by successive exact minimizations along such vectors of the convex quadratic function (6.29). The theorem follows from well-known properties of exact minimization along A-conjugate vectors (see for instance Luenberger (1973)). Q.E.D.

Remark 6.6
The identities (6.33) and (6.34) are valid whenever the implicit LU factorization algorithm is defined, say whenever A is strongly nonsingular. In such a case,

Theorems 6.9 and 6.10 can be reformulated by substituting the word 'minimizes' by the phrase 'makes stationary'.

6.4   DERIVATION OF THE PARAMETERS THAT GENERATE AN IMPLICIT LL^T FACTORIZATION

In this section we discuss some parameter choices that generate an implicit LL^T or Cholesky factorization, when the coefficient matrix is symmetric positive definite. The choices are defined in terms of a sufficient condition on the parameter z_i that gives freedom to the choices of H_1 and w_i. Thus a class of algorithms generating an implicit LL^T factorization is obtained. We show that, by suitably scaling the search vector in the implicit LU factorization algorithm, we can satisfy the sufficient condition for the implicit LL^T factorization. Thus the LU and LL^T factorizations can be implicitly obtained by the same algorithm.

Theorem 6.11

Let A be a symmetric positive definite matrix and let the sequence H_i be defined by (3.5) with admissible choices for H_1 and w_i. Let K_i be defined as in Lemma 6.1. For $i = 1, \ldots, n$, define the matrix T_i through the following recursion, starting with $T_0 = 0$:

$$T_i = T_{i-1} + H_i^T z_i z_i^T H_i . \tag{6.36}$$

Then for $i = 1, \ldots, n$ it is possible to choose z_i so that the following relations are true:

$$T_i a_j = e_j , \qquad j \leq i , \tag{6.37}$$

$$z_i^T H_i a_i \neq 0 \tag{6.38}$$

$$T_i = K_i. \tag{6.39}$$

Proof

We proceed by induction. For $i = 1$, (6.37) reads

$$(z_1^T H_1 a_1) H_1^T z_1 = e_1 . \tag{6.40}$$

Since H_1 is nonsingular, (6.40) can be solved for z_1 provided that $z_1^T H_1 a_1 \neq 0$. Premultiplying by a_1, we get

$$(z_1^T H_1 a_1)^2 = A_{1,1} \tag{6.41}$$

so that $z_1^T H_1 a_1$ is nonzero (and real) since $A_{1,1}$ is positive. Using (6.41), system (6.40) becomes

$$H_1^T z_1 = \frac{\pm e_1}{(A_{1,1})^{1/2}} . \tag{6.42}$$

Hence

$$T_1 = \begin{bmatrix} 1/A_{1,1} & 0 \\ 0 & 0 \end{bmatrix} \qquad (6.43)$$

or $T_1 = K_1$ and the theorem is true for $i = 1$. We assume now that properties (6.37)–(6.39) are true up to the index i. For $i + 1$ equation (6.37) reads

$$(z_{i+1}{}^T H_{i+1} a_{i+1}) H_{i+1}{}^T z_{i+1} = e_{i+1} - T_i a_{i+1} \qquad (6.44)$$

and is solvable if and only if $z_{i+1}{}^T H_{i+1} a_{i+1}$ is nonzero and $e_{i+1} - T_i a_{i+1}$ lies in the range of $H_{i+1}{}^T$. Multiplying on the left by a_{i+1} and noting that by induction $T_i = K_i$, (6.44) gives

$$(z_{i+1}{}^T H_{i+1} a_{i+1})^2 = a_{i+1}{}^T e_{i+1} - a_{i+1}{}^T K_i a_{i+1} \ . \qquad (6.45)$$

By Lemma 6.1 the right-hand side in (6.45) is strictly positive, since A is positive definite; hence $z_{i+1}{}^T H_{i+1} a_{i+1}$ is nonzero (and real). Now the range of $H_{i+1}{}^T$ is the orthogonal complement of the null space of H_{i+1}, which is spanned by the columns of A^i. Thus $e_{i+1} - T_i a_{i+1} = e_{i+1} - K_i a_{i+1}$ lies in the range of $H_{i+1}{}^T$ if and only if

$$(A^i)^T (e_{i+1} - K_i a_{i+1}) = 0 \ . \qquad (6.46)$$

From the symmetry of A we have

$$(A^i)^T e_{i+1} = (I_i, \ 0) a_{i+1} \qquad (6.47)$$

where I_i is the unit matrix in R^i. From the symmetry of K_i and the induction we have $(A^i)^T K_i = (K_i^T A^i)^T = (K_i A^i)^T = (I_i, \ 0)$, implying that (6.46) is identically satisfied. Multiplying now (6.44) on the left by e_j with $j > i + 1$ we obtain, since $T_i = K_i$,

$$e_j{}^T H_{i+1}{}^T z_{i+1} = 0 \ , \qquad (6.48)$$

implying that the matrix $H_{i+1}{}^T z_{i+1} z_{i+1}{}^T H_{i+1}$ and hence T_{i+1} has the form

$$T_{i+1} = \begin{bmatrix} B_{i+1} & 0 \\ 0 & 0 \end{bmatrix} \qquad (6.49)$$

with B_{i+1} square $i + 1$ by $i + 1$. Now (6.37) for $i + 1$ can be written compactly as

$$T_{i+1}A^{i+1} = \begin{bmatrix} I_{i+1} \\ 0 \end{bmatrix} .$$ (6.50)

Substituting (6.49) in (6.50) gives

$$B_{i+1}A^{i+1} = I_{i+1}$$ (6.51)

so that $T_{i+1} = K_{i+1}$, completing the induction. Q.E.D.

Theorem 6.12
Let A be symmetric positive definite and consider the algorithm of the ABS class with parameters H_1, w_i, z_i chosen as in Theorem 6.11. Then the associated implicit factorization is of the LL^T type.

Proof
For $i = n$ (6.39) can be written as

$$\sum_{j=1}^{n} p_j p_j^T = A^{-1}$$ (6.52)

or $A^{-1} = P^n(P^n)^T$. From Theorem 3.12 we have that $A^{-1} = P^n(L^n)^{-1}$, so that $P^n = [(L^n)^{-1}]^T$ is upper triangular. The theorem follows with $L^T = (P^n)^{-1}$. Q.E.D.

Remark 6.7
From the previous relations we obtain the following expression for the search vector (K_0 being the null matrix):

$$p_i = \beta_i(e_i - K_{i-1}a_i),$$ (6.53)

where β_i is a scalar defined by

$$\beta_i = \frac{\pm 1}{(a_i^T e_i - a_i^T K_{i-1}a_i)^{1/2}} .$$ (6.54)

Thus the search vector is uniquely defined, apart from a sign, independently of the choices for H_1 and w_i. Since the matrix P_n equals the inverse of the upper triangular factor in the implicit Cholesky factorization, the above characterization of the search vector reflects the well-known property that the Cholesky factorization is uniquely defined apart from n sign choices.

The following theorem gives another characterization of the search vectors generated by the implicit Cholesky factorization algorithm.

Theorem 6.13
The vectors $p_i = H_i^T z_i$ defined by the ABS algorithm with A, H_1, w_i, z_i as in Theorem 6.11 are A-conjugate.

Proof
Since A is symmetric, it is enough to prove that $p_j^T A p_i = 0$ for $1 \leqslant j < i$, $2 \leqslant i \leqslant n$. From (6.53) and (6.54) we have

$$Ap_i = \beta_i(a_i - AK_{i-1}a_i) \ . \tag{6.55}$$

Considering the partition of A given by

$$A = \begin{bmatrix} A^{i-1,i-1} & B_i \\ C_i & D_i \end{bmatrix} , \tag{6.56}$$

we obtain

$$AK_{i-1} = \begin{bmatrix} I_{i-1} & 0 \\ G_i & 0 \end{bmatrix} , \tag{6.57}$$

with $G_i = C_i(A^{i-1,i-1})^{-1}$. Thus we can write, since $\beta_j \beta_i$ is nonzero,

$$\frac{p_j^T A p_i}{\beta_j \beta_i} = e_j^T a_i - \left(\begin{bmatrix} I_{i-1} & G_i^T \\ 0 & 0 \end{bmatrix} e_j \right)^T a_i$$

$$- a_j^T K_{j-1} a_i + a_j^T K_{j-1} \begin{bmatrix} I_{i-1} & 0 \\ G_i & 0 \end{bmatrix} a_i \ . \tag{6.58}$$

Since $j < i$, we get identically

$$\begin{bmatrix} I_{i-1} & G_i^T \\ 0 & 0 \end{bmatrix} e_j = e_j \tag{6.59}$$

and

$$K_{j-1} \begin{bmatrix} I_{i-1} & 0 \\ G_i & 0 \end{bmatrix} = K_{j-1} \ . \tag{6.60}$$

Hence the right-hand side in (6.58) is identically zero. Q.E.D.
    The implicit Cholesky algorithm could be implemented using (6.53) and (6.54)

for the search vector, with K_{i-1} updated according to (6.36) and no restriction on the choice of H_i. The following theorem shows, however, that essentially the same parameter choices of the implicit LU factorization algorithm satisfy the condition of Theorem 6.11 when A is symmetric positive definite, the only modification being the introduction of a scaling factor for z_i (or for p_i).

Theorem 6.14
Let A be symmetric positive definite. Then the algorithm of the ABS class defined by the choices $H_1 = I$, $w_i = e_i/e_i^T H_i a_i$, $z_i = \beta_i e_i$, $\beta_i = \pm 1/(e_i^T H_i a_i)^{1/2}$ is well defined. Moreover for this choice of z_i equations (6.37) and (6.38) are satisfied.

Proof
From Theorem 6.5 the given choices for H_1, w_i are well defined and $e_i^T H_i a_i$ is positive from Corollary 6.1. The choice for z_i is therefore well defined and (6.38) is satisfied. Now (6.37) for $i = 1$ reads

$$\beta_1^2 a_1^T e_1 e_1 = e_1 \tag{6.61}$$

which is true from the definition of β_1. Assuming that (6.37) is true up to the index i, we have to show that $u_{i+1} = v_{i+1}$, where

$$u_{i+1} = \beta_{i+1}^2 H_{i+1}^T e_{i+1} e_{i+1}^T H_{i+1} a_{i+1} \tag{6.62}$$

and

$$v_{i+1} = e_{i+1} - K_i a_{i+1} . \tag{6.63}$$

The identity of u_{i+1} and v_{i+1} is conveniently proved by establishing the identity of Ru_{i+1} and Rv_{i+1}, where R is the following nonsingular matrix:

$$R = \begin{bmatrix} a_1^T \\ \cdots \\ a_i^T \\ 0 \quad I_{n-i} \end{bmatrix} . \tag{6.64}$$

We show that $Ru_{i+1} = Rv_{i+1} = e_{i+1}$. For $j \leqslant i$ indeed we have

$$e_j^T R u_{i+1} = a_j^T H_{i+1}^T e_{i+1} \tag{6.65}$$

which is zero since a_j lies in the null space of H_{i+1}. Also we have

$$e_j^T R \upsilon_{i+1} = a_j^T e_{i+1} - e_j^T a_{i+1} \tag{6.66}$$

which is zero because of symmetry of A. For $j > i + 1$ we have

$$e_j^T R u_{i+1} = e_j^T H_{i+1}^T e_{i+1} \tag{6.67}$$

which is zero owing to the structure of H_{i+1}, see (6.10). Also we have

$$e_j^T R \upsilon_{i+1} = e_j^T e_{i+1} + e_j^T K_i a_{i+1} \tag{6.68}$$

which is zero since $K_i^T e_j = K_i e_j$ is identically zero. Now the $(i+1)$th component satisfies $e_{i+1}^T R \upsilon_{i+1} = 1$ since $K_i^T e_{i+1} = K_i e_{i+1} = 0$; moreover $e_{i+1}^T R u_{i+1} = e_{i+1}^T H_{i+1} e_{i+1}$ is equal to one for $H_1 = I$ from the structure of H_{i+1}, see (6.10). Q.E.D.

Remark 6.8
If we define p_i by

$$p_i = \frac{H_i^T e_i}{(|e_i^T H_i a_i|)^{1/2}} , \tag{6.69}$$

then from Theorem 6.14 the resulting implicit *LU* factorization algorithm is well defined, satisfies the conditions of Theorem 6.11 and therefore $U = L^T$.

Definition 6.4
The implicit Gauss–Cholesky or the implicit LU–LL^T factorization algorithm is the algorithm of the ABS class with the following parameter choices: $H_1 = I$, $w_i = e_i/e_i^T H_i a_i$, $z_i = e_i/(|e_i^T H_i a_i|)^{1/2}$.

It does not appear that use can be made of the symmetry of A to reduce the computational cost of the implicit Gauss–Cholesky algorithm from $n^3/3 + O(n^2)$ multiplications to $n^3/6 + O(n^2)$, as is the case for the classical Cholesky algorithm. However, we show in section 6.7 that the additional cost for inverting a positive definite matrix with the implicit Gauss–Cholesky algorithm is half the cost required by the classical procedure, i.e. it is $n^3/6 + O(n^2)$, hence the total cost for the inversion is the same in the classical and in the ABS approach.

6.5 ALTERNATIVE REPRESENTATION OF THE IMPLICIT *LU* FACTORIZATION ALGORITHM

We consider in this section the application to the implicit *LU* factorization algorithm of the various alternative formulations considered in Chapter 4.

Let us first consider (4.6). The expression for W^i is the following:

$$W^i = \begin{bmatrix} D_i \\ 0 \end{bmatrix} , \tag{6.70}$$

where D_i is the i by i diagonal matrix whose jth diagonal element has the value $1/p_j^T a_j$. Therefore we obtain after straightforward algebra

$$H_{i+1} = I - \begin{bmatrix} 0 & 0 \\ \overline{A}^i(A^{i,1})^{-1} & I_i \end{bmatrix} , \tag{6.71}$$

where $\overline{A}^i$ is the matrix comprising the last $n - i$ rows of A^i. Note that a comparison of (6.71) with (6.10) gives the identity

$$S_i = -\overline{A}^i(A^{i,i})^{-1} . \tag{6.72}$$

Equation (6.71) is mainly of theoretical interest. Use of it has been made in the study of structured systems, including one related to the linear programming problem (Abaffy and Spedicato 1989).

In the use of (4.25) and (4.33) the following are the values of σ, q and r:

$$\sigma = \frac{e_i^T a_i}{p_i^T a_i} , \tag{6.73}$$

$$q = D_i a_i' , \tag{6.74}$$

$$r = \frac{\overline{a}_i'}{p_i^T a_i} , \tag{6.75}$$

where a_i', $\overline{a}_i'$ are respectively the vectors in R^i comprising the first i components of the ith row and of the ith column of A. If we consider equations (4.37)–(4.39), we see that the vectors p_i and s_i are proportional:

$$p_i = (p_i^T a_i)s_i \tag{6.76}$$

so that only the two sequences p_i and r_i need to be stored. They are defined by relations $p_1 = e_1$, $r_1 = a_1$ and

$$p_i = e_i - \sum_{j=1}^{i-1} \frac{r_j^T e_i}{p_j^T a_j} p_j , \tag{6.77}$$

$$r_i = a_i - \sum_{j=1}^{i-1} \frac{p_j^\mathrm{T} a_i}{p_j^\mathrm{T} a_j} r_j \ . \tag{6.78}$$

The above formulas are more expensive than the standard ones both in memory requirement (even if underdetermined systems are solved) and in computational cost (they require $(5/6)n^3 + O(n^2)$ multiplications for a square system).

Let us consider (4.42) and (4.43). Since by (4.44) u_j^i equals the jth row of H_i, it follows from (6.10) that u_j^i has only i possible nonzero elements, precisely the first $i-1$ elements and the jth element which is equal to one.

It is seen that $3(i-1)$ multiplications and one division are needed in (4.42), while $2in\text{-}2i^2 + O(n)$ multiplications are needed in (4.43); thus the total cost is $n^3/3 + O(n^2)$ multiplications as in the standard implementation. The memory occupation is similarly $n^2/4 + O(n)$.

We conclude this section by considering the modification of the implicit LU factorization algorithm where at the $(i+1)$th iteration the special step is taken which satisfies the whole system. In the case of the modified Huang algorithm this approach has led to an algorithm (the fast modified Huang algorithm) which requires fewer multiplications than the standard modified Huang algorithm. In the present case we find that no similar improvement can be obtained.

The equation governing the special step is (3.52). Since the first i rows of H_{i+1} are null, $Z = (H_{i+1}A^{n-i})^\mathrm{T}$ has the form

$$Z = (0, \ Z'^\mathrm{T}) \tag{6.79}$$

so that system (3.52) reads

$$(0, \ Z'^\mathrm{T}) \begin{bmatrix} s_1 \\ s_2 \end{bmatrix} = \begin{bmatrix} 0 \\ u_2 \end{bmatrix} \ , \tag{6.80}$$

with Z' a square nonsingular matrix of dimension $n-i$. Thus s_2 is uniquely defined by

$$Z'^\mathrm{T} s_2 = u_2 \ . \tag{6.81}$$

Setting $s_1 = 0$ gives a particular solution of (6.80) which is both a basic type solution and a minimal Euclidean norm solution.

We now consider the total cost of solving the original system up to the ith equation, of constructing Z' and solving (6.81) using the implicit LU factorization algorithm. In solving the first i equations, $2j(n-j)$ multiplications are required at the jth step, so that the total cost is $ni^2 - (2/3)i^3$. In the computation of each column of Z', $(n-i)i$ multiplications are required, for a total cost of $(n-i)^2 i$ multiplications. In solving (6.81), $n(n-i)^2 - (2/3)(n-i)^3$ multiplications are required. Thus the total number $\Phi(n, i)$ is given by

$$\Phi(n, i) = i^3 - 2ni^2 + n^2i + \frac{n^3}{3} . \tag{6.82}$$

Now $\Phi(n, 0) = n^3/3$, $\Phi(n, n) = n^3/3$, $(\delta\Phi/\delta i)_{i=0} = n^2$, $(\delta\Phi/\delta i)_{i=n} = 0$ implying, since Φ has degree three, that $\Phi(n, i) > n^3/3$ for $0 < i < n$. Actually, the stationary points of Φ are at $i = n/3$, a maximum with $\Phi(n, n/3) = (13/27)n^3$, and at $i = n$, a minimum. Thus no improvement in terms of number of multiplications can be obtained by using the special iteration.

6.6   RELATIONSHIPS WITH THE ESCALATOR METHOD, THE METHOD OF BROWN AND THE METHOD OF SLOBODA

In the classical escalator method, considered in Chapter 2, the solution x^+ of $Ax = b$ is approximated by a sequence of iterates $\bar{x}_1, \ldots, \bar{x}_n = x^+$ whose ith member has the form

$$\bar{x}_I \equiv \begin{bmatrix} (A^{i,i})^{-1}b^i \\ 0 \end{bmatrix} , \tag{6.83}$$

where b^i is the vector in R^i comprising the first i components of b. Clearly the escalator method is well defined (without pivoting) if and only if A is strongly nonsingular. If the implicit LU factorization algorithm is considered with $x_1 = 0$, then, from Theorem 6.8, the last $n - i$ components of x_{i+1} are zero. Since x_{i+1} solves the first i equations, it follows, if $A^{i,i}$ is nonsingular, that it must be of the form (6.83). Thus the identity $\bar{x}_i = x_{i+1}$ holds, implying that the sequence of approximations to the solution generated by the escalator method is a subclass of the sequences generated by the implicit LU factorization algorithm (with respect to the possible choices of x_1).

The Brown (1969) algorithm has been introduced as an algorithm for nonlinear equations based upon Gaussian elimination. It consists of cycles, each one having n steps and solving a certain linear equation. The original algorithm of Brown is rather complex and requires $O(n^4)$ multiplications per cycle. It has been modified by Gay (1975a,b), who has given a version requiring only $n^3/3 + O(n^2)$ multiplications per cycle. Here we consider the Gay method as presented by Moré and Cosnard (1979).

Let $Ax = b$ be the linear system to be solved in a cycle with A nonsingular. In the Brown method, as in the Brent method, at the ith step of the cycle an approximation $\bar{x}_{i+1}$ is obtained which solves the first i equations. Writing $\bar{x}_{i+1}$ as $\bar{x}_i + \bar{p}_i$, $\bar{p}_i$ is a solution of the following underdetermined system, with δ_{ij} the Kronecker delta:

$$a_j^T\bar{p}_i = -\delta_{ij}(a_i^T\bar{x}_i - b^Te_i) , \qquad 1 \leqslant j \leqslant i . \tag{6.84}$$

In the Brown method a solution of (6.84) is sought which has at least $n - i$ zero components. It is determined as follows. Suppose that a nonsingular n by n matrix R_{i+1} exists such that

$$A^{iT}R_{i+1} = (L_i, 0) \ ,$$ (6.85)

where L_i is an i by i nonsingular lower triangular matrix with diagonal elements σ_1, ..., σ_i. Writing (6.84) in the form

$$(a_j^T R_{i+1})(R_{i+1})^{-1}\bar{p}_i = -\delta_{ij}(a_i^T \bar{x}_i - b^T e_i) \ , \quad 1 \leqslant j \leqslant i \ ,$$ (6.86)

it follows from (6.86) that

$$(R_{i+1})^{-1}\bar{p}_i = \sum_{j=1}^{n} \theta_j e_j \ ,$$ (6.87)

with θ_j arbitrary for $j > i$ and $\theta_i = -(a_i^T x_i - b^T e_i)/\sigma_i$. It can now be shown, (Moré and Cosnard 1979) that, if R_1 is an arbitrary nonsingular matrix, then, for $j = 1, \ldots, i$, elementary permutation matrices P_j and elementary upper triangular matrices T_j exist such that (6.85) is satisfied (with j replacing i). Moreover the following relation is true for R_{i+1}

$$P_i \ldots P_1 R_{i+1} = \begin{bmatrix} U_i & S_i \\ 0 & I_{n-i} \end{bmatrix} \ ,$$ (6.88)

where U_i is unit upper triangular of order i. Equations (6.87) and (6.88) imply that the vector $P_i \ldots P_1 \bar{p}_i$ has the last $n - i$ components equal to $\theta_{i+1}, \ldots, \theta_n$. Thus, taking all the θ_j equal to zero, gives to $\bar{p}_i$ the required zero structure. When A is strongly nonsingular, the matrix P_i can be taken as the unit matrix, so that the last $n - i$ components of $\bar{p}_i$ are zero. In such a case it is clear that, if the Brown method and the implicit LU factorization algorithm are started with the same vectors $\bar{x}_1 = x_1$, then the sequences $\bar{x}_i$, x_i are identical. In general, however, such sequences may differ, owing to different pivoting strategies.

Finally we consider the algorithm described by Sloboda (1978) for linear systems with a strongly nonsingular matrix A. This algorithm is based upon the recursions for $i = 1, \ldots, n$ and $j = i + 1, \ldots, n + 1$:

$$x_{i+1}^j = x_i^j - \frac{a_i^T x_i^j - b^T e_i}{(a_i^T v_i^i)} v_i^i,$$ (6.89)

$$v_{i+1}^{i+1} = x_{i+1}^{i+2} - x_{i+1}^{i+1},$$ (6.90)

started with x_1^1 arbitrary, $x_1^{1+j} = x_1^1 + e_j, j = 1, \ldots, n$ and $v_1^1 = e_1$. Two properties of the above iterates are the following.

— The vectors x_{i+1}^j satisfy the first i equations.
— The left A-semiconjugacy property $v_j^T A v_i^i = 0, j < i$ is satisfied.

If we define the vectors v_i^j by

$$v_i^j = x_i^{j+1} - x_i^i , \qquad j = 1, \ldots, n , \tag{6.91}$$

then the following recursion is satisfied:

$$v_{i+1}^j = v_i^j - \frac{a_i^T v_i^j}{(a_i^T v_i^i)} v_i^i , \tag{6.92}$$

which has the same form as the recursion (4.42) with v_i^j substituting u_i^j. For the implicit LU factorization algorithm $u_j^1 = e_j$, implying that the vector $p_i = u_i^i$ in this algorithm is identical with the vector v_i^i in the Sloboda algorithm. Thus, if $x_1^{n+1} = x_1$, the vector x_i generated by the implicit LU factorization algorithm is identical with the vector x_i^{n+1} generated by Sloboda algorithm. The two algorithms also share a similar storage, $n^2/4 + O(n)$, and computational cost, $n^3/3 + O(n^2)$ multiplications. We note, however, that the Sloboda algorithm generates at the ith step $n + 1 - i$ approximations of the solution and that no use of a_i is made in the evaluation of the search vector v_i^i. This is a possible advantage, particularly in the context of parallel computation and in the applications to nonlinear equations and nonlinear optimization.

6.7   COMPUTING THE INVERSE BY THE IMPLICIT LU FACTORIZATION ALGORITHM

An explicit formula for the inverse of a symmetric strongly nonsingular matrix can be given in terms of the search vectors generated by the implicit LU factorization algorithm. This is given in Theorem 6.15 where use is made of the following Lemma 6.2, due to Egervary (1960) as a generalization of a previous result of Hestenes and Stiefel (1952) for square symmetric positive definite matrices.

Lemma 6.2
Let A be a matrix of dimension m by n and of rank r. Let $u_1, \ldots, u_r \in R^n$ and $v_1, \ldots, v_r \in R^m$ be such that $v_i^T A u_j = 0$ for $i \neq j$, while $v_i^T A u_i \neq 0$ (say the vectors v_i, u_i are A-biconjugate). Then the matrix A^+ given by

$$A^+ = \sum_{i=1}^r \frac{u_i v_i^T}{v_i^T A u_i} \tag{6.93}$$

is a generalized inverse of A.

Theorem 6.15

Let A be symmetric strongly nonsingular. Let $p_1, \ldots, p_n$ be the search vectors generated by the implicit LU factorization algorithm. Then the following dyadic expansion of the inverse A^{-1} is true:

$$A^{-1} = \sum_{i=1}^{n} \frac{p_i p_i^{\mathsf{T}}}{a_i^{\mathsf{T}} p_i} . \tag{6.94}$$

Proof

From Corollary 6.4 the vectors p_i are A-conjugate; so we can apply Lemma 6.2 with $u_i = v_i = p_i$. Equation (6.94) follows using the identity (6.34). Q.E.D.

Remark 6.9

If the vectors p_i are stored, the computation of the inverse of a symmetric strongly nonsingular matrix via (6.94) requires $n^3/6 + O(n^2)$ additional multiplications (for a total value, including computation of the p_i, of $n^3/2 + O(n_2)$ multiplications).

If only k terms are used in (6.94), then the resulting dyadic expansion is related to the inverse of the kth principal submatrix of A, as shown in the following.

Theorem 6.16

Let $p_1, \ldots, p_k$ be the first k search vectors $(k \le n)$ generated by the implicit LU factorization algorithm on the symmetric strongly nonsingular matrix A. Then the following identity is true:

$$\sum_{i=1}^{k} \frac{p_i p_i^{\mathsf{T}}}{a_i^{\mathsf{T}} p_i} = \begin{bmatrix} (A^{k,k})^{-1} & 0 \\ 0 & 0 \end{bmatrix} . \tag{6.95}$$

Proof

Since the vectors p_i are the columns of an upper triangular matrix, we have the identity

$$\sum_{i=1}^{k} \frac{p_i p_i^{\mathsf{T}}}{a_i^{\mathsf{T}} p_i} = \begin{bmatrix} B_k & 0 \\ 0 & 0 \end{bmatrix} , \tag{6.96}$$

with B_k some matrix. It is therefore enough to prove that

$$\sum_{i=0}^{k} \frac{p_i p_i^{\mathsf{T}}}{a_i^{\mathsf{T}} p_i} \begin{bmatrix} A^{k,k} & 0 \\ 0 & 0 \end{bmatrix} = \begin{bmatrix} I_k & 0 \\ 0 & 0 \end{bmatrix} . \tag{6.97}$$

Since the last $n - k$ components of p_i are zero, we have the identity

$$p_i^T \begin{bmatrix} A^{k,k} & 0 \\ 0 & 0 \end{bmatrix} = p_i^T(A^k, 0) . \tag{6.98}$$

Now $p_i^T(A^k, 0) = e_i^T H_i(A^k, 0) = e_i^T(H_i A^k, 0) = e_i^T(0, \ldots, 0, H_i a_i, H_i a_{i+1}, \ldots, H_i a_k, 0, \ldots, 0) = u_i^T$, where

$$u_i^T = (0, \ldots, 0, a_i^T p_i, a_i^T H_i a_{i+1}, \ldots, a_i^T H_i a_k, 0, \ldots, 0) . \tag{6.99}$$

From (6.99) it follows that $p_i u_i^T$ is a matrix whose elements are all zero, save that on the intersection of the ith row with the ith column, which is equal to $a_i^T p_i$. It follows therefore that each matrix in the summation in (6.97) has all the elements zero save that on the intersection of the ith row and the ith column, which is equal to one. Thus (6.97) is true, proving the theorem. Q.E.D.

For nonsymmetric matrices we can apply the general procedure outlined in Chapter 3 for the computation of the inverse. Owing to the particular properties of the implicit LU factorization algorithm, a further reduction in the computational cost can be obtained, as shown in the next theorem, giving a total cost which is the same as for the classical procedure.

Theorem 6.17
The inverse of a strongly nonsingular matrix can be computed in the framework of the implicit LU factorization algorithm in no more than $n^3 + O(n^2)$ multiplications.

Proof
We consider the procedure given in section 3.5 using the zero vector as starting point and the vectors p_i generated by the implicit LU factorization algorithm. Since such vectors are the columns of an upper triangular matrix and since, from Theorem 6.8 and the assumption on the starting point, the last $n - k$ components of the $(k + 1)$th approximation $(x_i)_{k+1}$ to the solution of the ith system are zero, it follows that only $2k$ multiplications are needed for the update of $(x_i)_{k+1}$. Thus the number of multiplications required for solving the ith system is no more than $\sum_{i=1}^{n} (2k) = n^2 - i^2$, giving for all the systems a total amount $(2/3)n^3 + O(n^2)$. The theorem follows since the computation of the search vectors can be done in $n^3/3 + O(n^2)$ multiplications. Q.E.D.

6.8   BIBLIOGRAPHICAL REMARKS

Almost all the results in this chapter are due to Broyden (Theorems 6.1, 6.2, 6.11 and 6.12 and Lemma 6.1) and to Spedicato (the remaining theorems). Some of the results have appeared, sometimes with different proofs, in the paper by Abaffy *et al.* (1982) and, without proof, in the paper by Abaffy *et al.* (1984a).

7

The scaled ABS algorithm: general formulation

7.1 INTRODUCTION

In this chapter we consider an essential generalization of the ABS algorithm obtained by the introduction at each iteration of a new parameter vector, called the scaling vector. The basic ABS algorithm introduced in the previous chapters will also be called in the following the unscaled ABS algorithm (since it corresponds to taking the scaling vector as the unit vector).

The scaled ABS algorithm is formally and computationally different, but equivalent, in the sense of generating the same set of iterates x_i, to some general procedures for linear systems that can be traced back at least to Householder (1955) and that have been developed to some extent by Stewart (1973) and Broyden (1985). The relationship with the Broyden procedure shows that a fundamental characterization of the scaled ABS algorithm is that it contains all possible algorithms of a very general iterative form that find the solution, from an arbitrary starting point, in a number of steps no greater than the number of equations.

In this chapter we consider the basic properties of the scaled ABS algorithm, many of them being immediate extensions of the corresponding properties in the unscaled class. Among the properties not previously considered, we give a block formulation of the algorithm. The analysis of special properties corresponding to particular choices of the scaling vector is presented in the following chapter.

7.2 DERIVATION OF THE SCALED ABS ALGORITHM

Let us consider, instead of the original system $Ax = b$, the following scaled system:

$$V^{\mathrm{T}}Ax = V^{\mathrm{T}}b, \tag{7.1}$$

where V is an arbitrary nonsingular m by m matrix. System (7.1) is equivalent to system $Ax = b$, any solution of one being a solution of the other. Note that, by letting $v_i \in R^m$ be the ith column of V, we can write system (7.1) componentwise in the form

$$v_i^T Ax = v_i^T b, \qquad i = 1, \ldots, m, \tag{7.2}$$

or also, by introducing the (unscaled) residual vector $r = r(x) \in R^m$,

$$r(x) = Ax - b \tag{7.3}$$

in the form

$$v_i^T r(x) = 0, \qquad i = 1, \ldots, m. \tag{7.4}$$

If the basic ABS algorithm is applied to solving the system (7.1), it is immediately seen that the relevant recursions are obtained by substituting a_i with $A^T v_i$ and $b^T e_i$ with $b^T v_i$. Thus we obtain the following.

ALGORITHM 5: The Scaled ABS Algorithm
(A5) Let $x_1 \in R^n$ be arbitrary; let $H_1 \in R^{n,n}$ be arbitrary nonsingular; set $i = 1$ and $i$flag $= 0$.
(B5) Let $v_i \in R^m$ be arbitrary save that $v_1, \ldots, v_i$ be linearly independent. Compute the residual vector r_i. If $r_i = 0$, stop, x_i solves the system. Otherwise compute the scalar $\tau_i = v_i^T r_i$ and the vector $s_i = H_i A^T v_i$.
(C5) If $s_i \neq 0$, go to (D5); if $s_i = 0$ and $\tau_i = 0$, set $x_{i+1} = x_i$, $H_{i+1} = H_i$, $i$flag $= i$flag $+ 1$ and go to (G5) if $i < m$; otherwise stop; if $s_i = 0$ and $\tau_i \neq 0$, set $i$flag $= -i$ and stop.
(D5) Compute the search vector p_i by

$$p_i = H_i^T z_i \tag{7.5}$$

where $z_i \in R^n$ is arbitrary save for the condition

$$z_i^T H_i A^T v_i \neq 0. \tag{7.6}$$

(E5) Update the approximation of the solution by

$$x_{i+1} = x_i - \alpha_i p_i \tag{7.7}$$

where the step size α_i is given by

$$\alpha_i = \frac{\tau_i}{v_i^T A p_i} \tag{7.8}$$

If $i = m$, stop, x_{m+1} solves the system.
(F5) Update the matrix H_i by

$$H_{i+1} = H_i - H_i A^T v_i w_i^T H_i \tag{7.9}$$

where $w_i \in R^n$ is arbitrary save for the condition

$$w_i^T H_i A^T v_i = 1. \tag{7.10}$$

(G5) Increment the index i by one and go to (B5).

We make some simple but fundamental observations on ALGORITHM 5.

— If $V = I$, i.e. $v_i = e_i$, then ALGORITHM 5 is identical with ALGORITHM 1.
— Since system (7.1) is equivalent to system $Ax = b$, the vector x_{m+1} solves $Ax = b$. Thus ALGORITHM 5 can be interpreted, for general V, as an algorithm for solving $Ax = b$.
— The vector v_i plays the role of an additional parameter available at the ith iteration, arbitrary, save that $v_1, \ldots, v_i$ must be linearly independent.

Definition 7.1
The matrix V in (7.1) is called the scaling matrix. The columns of V are called the scaling vectors.

Definition 7.2
The class of algorithms for solving the system $Ax = b$ defined by ALGORITHM 5 is called the scaled ABS class of algorithms. The subclass of the scaled ABS class where at each step $v_i = e_i$ is called the basic or unscaled ABS class of algorithms.

7.3 BASIC PROPERTIES OF THE SCALED ABS CLASS

Since every scaled ABS algorithm is obtained by applying an unscaled ABS algorithm to the scaled equations, it follows that every property of the unscaled ABS algorithm of the form $Q(H_i, a_i, b_i, z_i, w_i)$ can be reformulated as a property of the scaled ABS algorithm of the form $Q(H_i, A^T v_i, b^T v_i, z_i, w_i)$. The following theorems can therefore be obtained by immediate reformulation of the previously proved theorems and the details of their proofs can be omitted. For simplicity of formulation we shall also assume, unless otherwise stated, that A is full rank.

Theorem 7.1
For $1 \leq i \leq j \leq m$ the vectors $H_i A^T v_j$ are nonzero and linearly independent.

Theorem 7.2
For $1 \leq i \leq m$ the vectors $A^T v_1, \ldots, A^T v_i$ are nonzero, linearly independent and a basis of the null space of H_{i+1}. The vectors $w_1, \ldots, w_i$ are nonzero, linearly independent and a basis of the null space of H_{i+1}^T.

Remark 7.1

If we define the full rank matrix V^i by

$$V^i = (v_1, \ldots, v_i) \tag{7.11}$$

then Theorem 7.2 implies that

$$H_{i+1}A^T V^i = 0. \tag{7.12}$$

Theorem 7.3

Define the matrix $\overline{W}_i$ as in (3.33) and the matrix $\overline{A}^i$ by

$$\overline{A}^i = (H_1 A^T v_1, \ldots, H_i A^T v_i). \tag{7.13}$$

Then the matrices $\overline{W}^i$ and $\overline{A}^i$ are full rank. Moreover, if H_1 is symmetric, the columns of $\overline{W}^i$ and $\overline{A}_i$ are mutually H_1^{-1}-conjugate. If $H_1 = I$ and $m = n$, then $\overline{W}^n = (\overline{A}^n)^{-1}$.

Theorem 7.4

Define P^i as in (3.37) and define the matrix L^i by

$$L^i = (V^i)^T A P^i. \tag{7.14}$$

Then the matrix L^i is lower triangular and nonsingular. Hence the search vectors are linearly independent and the following implicit factorization holds, with $P = P^m$, $L = L^m$:

$$V^T A P = L. \tag{7.15}$$

Remark 7.2

If A is nonsingular, Theorem 7.4 can be proved with the only condition on V_i that (7.6) and (7.10) are satisfied. The linear independence of the scaling vectors follows from (7.14).

Theorem 7.5

The residual vector evaluated in x_{i+1} is orthogonal to the first i columns of V, i.e.

$$V^{iT} r_{i+1} = 0. \tag{7.16}$$

Any vector $\bar{x}$ such that the residual $r(\bar{x})$ is orthogonal to the first i columns of V has the form

$$\bar{x} = x_{i+1} + H_{i+1}^T s \tag{7.17}$$

for s some vector in R^n. Moreover any such vector $\bar{x}$ can be obtained at the ith step of the scaled ABS as x_{i+1} by suitable choice of z_i.

Remark 7.3

The termination property of the scaled ABS algorithm after m steps follows immediately from Theorem 7.5, since the m-dimensional vector $r(x_{m+1})$ is orthogonal to m linearly independent vectors.

7.4 ALTERNATIVE FORMULATIONS OF THE SCALED ABS ALGORITHM

The next theorems give alternative formulations of the matrix H_i and the search vectors p_i, by generalization of similar formulations given in Chapter 4.

Theorem 7.6

Define $\overline{W}^i$ as in (3.33) and $\overline{A}^i$ as in (7.13). Then the matrix H_{i+1} can be written in the form

$$H_{i+1} = H_1 - \overline{A}^i(\overline{W}^i)^{\mathrm{T}}. \tag{7.18}$$

Theorem 7.7

Define the matrix $Q^i \in R^{i,i}$ by

$$Q^i = (W^i)^{\mathrm{T}} H_1 A^{\mathrm{T}} V^i. \tag{7.19}$$

Then the matrix Q^i is strongly nonsingular and the matrix H_{i+1} can be written in the form

$$H_{i+1} = H_1 - H_1 A^{\mathrm{T}} V^i (Q^i)^{-1} (W^i)^{\mathrm{T}} H_1 \tag{7.20}$$

Theorem 7.8

The search vectors p_i generated by the scaled ABS algorithm can be written using equation (4.37), where s_j, u_j are vectors in R^n given by $s_1 = H_1 A^{\mathrm{T}} v_1$, $u_1 = H_1^{\mathrm{T}} w_1$ and, for $i > 1$, u_i is given by (4.39) and s_i by

$$s_i = H_1 A^{\mathrm{T}} v_i - \sum_{j=1}^{i-1} s_j u_j^{\mathrm{T}} A^{\mathrm{T}} v_i \tag{7.21}$$

Theorem 7.9

Let $z_1, \ldots, z_m$ be a set of H_1-admissible vectors in step (D5) of the scaled ABS algorithm. Define the vector w_i by

$$w_i = \frac{z_i}{z_i^{\mathrm{T}} H_i A^{\mathrm{T}} v_i} \tag{7.22}$$

Consider for $i = 1, \ldots, m$ the $m - i$ vectors $u_j^{i+1} \in R^n$ defined by the recursion

$$u_j^{i+1} = u_j^i - \frac{v_i^T A u_j^i}{v_i^T A u_i^i} u_i^i \quad , j = i+1, \ldots, m, \tag{7.23}$$

with

$$u_j^1 = H_1^T z_j, \qquad j = 1, \ldots, m \tag{7.24}$$

Then, for $j = 1, \ldots, m$, we have the identity $H_i^T z_j = u_i^i$, implying that $p_i = u_i^i$. Moreover if $m = n$ the set of vectors $\bar{x}$ such that the residual $r(\bar{x})$ is orthogonal to v_1, $\ldots, v_i$ consists of the vectors of the form (4.47), where x_{i+1} is the approximation of the solution obtained at the ith step of the scaled ABS algorithm and $d \in R^{n-i}$ is arbitrary.

Equations (4.50) for $p_i = p_i^i$ become now

$$p_i^{j+1} = p_i^j - \frac{v_j^T A p_i^j}{v_j^T A p_j} p_j, \qquad j = 1, \ldots, i-1, \tag{7.25}$$

starting with $p_i^1 = z_i$. Now we have $v_j^T A p_i^j = v_j^T A H_j^T z_i = v_j^T A H_j^T H_j^T z_i = s_j^T p_i^j$; hence $A^T v_j$ can be replaced by $s_j = H_j A^T v_j$, once $s_1, \ldots, s_i$ are computed and stored. This leads to the recursions, for $j = 1, \ldots, i-1$,

$$p_i^{j+1} = p_i^j - \frac{s_j^T p_i^j}{s_j^T p_j} p_j, \tag{7.26}$$

$$s_i^{j+1} = s_i^j - \frac{p_j^T s_i^j}{p_j^T s_j} s_j, \tag{7.27}$$

with $s_i = s_i^i$. It must be assumed that z_i and v_i are such that $s_j^T p_j$ are nonzero for all j.

Remark 7.4
Equations (7.26) and (7.27) require $2mn$ storage location and $5m^2n/2$ multiplications.

7.5   BLOCK FORMULATION OF THE SCALED ABS ALGORITHM

In this section we introduce a block formulation of the scaled ABS algorithm. The partial Huang algorithm studied in Chapter 5 can be considered as a particular case of such a block formulation.

Definition 7.3
Let $V \in R^{m,m}$, $A \in R^{m,n}$ and $U \in R^{n,m}$ be full rank matrices ($m \leq n$). Define the matrix $L \in R^{m,m}$ by

$$L = V^T A U. \tag{7.28}$$

Then the pair (V, U) is called

(a) A-semiconjugate or A-semiorthogonal if L is lower triangular,
(b) regular A-semiconjugate or regular A-semiorthogonal if L is nonsingular lower triangular and
(c) A-conjugate or A-orthogonal if L is nonsingular diagonal.

Remark 7.5
The pair (V, U) of matrices whose columns are respectively the scaling and the search vectors of the scaled ABS algorithm is a regular A-semiconjugate pair.

Definition 7.4
Let V, A, U be given as in Definition 7.3. Let V and U be partitioned as follows:

$$V = (V_1, \ldots, V_p), \tag{7.29}$$

$$U = (U_1, \ldots, U_p), \tag{7.30}$$

where $1 \leqslant p \leqslant m$, $V_i \in R^{m, m_i}$, $U_i \in R^{n, m_i}$, $m_1 + \ldots + m_p = m$. Define the matrix $L_{ij} \in R^{m_i, m_j}$ by

$$L_{ij} = V_i^T A U_j \tag{7.31}$$

Then the pair (V, U) is called, with respect to the given partition,

(a) block A-semiconjugate or block A-semiorthogonal if $L_{ij} = 0$ for $i < j$ and $i = 1, \ldots, p - 1, j = i + 1, \ldots p$ (if moreover the p matrices L_{ii} are nonsingular, the pair (V, U) is called block regular A-semiconjugate or block regular A-semiorthogonal) and
(b) block A-conjugate or block A-orthogonal if $L_{ij} = 0$ for $i \neq j$ and moreover the p matrices L_{ii} are nonsingular.

The basic and the scaled ABS algorithms have been introduced by dealing, at the ith step, with just one equation. We can derive, by a similar approach, a generalization of these algorithms which deals, at the ith step, with one block of m_i equations.

Let $Ax = b$ be the system to be solved with $A \in R^{m, n}$ full rank. Let $V \in R^{m, m}$ be nonsingular and partitioned as in (7.29). Let $x_1 \in R^n$ be arbitrary and consider an iteration of the form

$$x_{i+1} = x_i - P_i d_i, \tag{7.32}$$

where P_i is n by m_i and $d_i \in R^{m_i}$. We want to determine P_i and d_i so that the residual r_{i+1} is orthogonal to the columns of $V_1, \ldots, V_i$, i.e.

$$V_j^T r_{i+1} = 0, \qquad j = 1, \ldots, i. \tag{7.33}$$

For $i = 1$, (7.33) gives, since $r_2 = r_1 - AP_1 d_1$,

$$V_1^{\mathrm{T}}r_1 - V_1^{\mathrm{T}}AP_1d_1 = 0. \tag{7.34}$$

Assuming that $V_1^{\mathrm{T}}AP_1$ is nonsingular, we get

$$d_1 = (V_1^{\mathrm{T}}AP_1)^{-1}V_1^{\mathrm{T}}r_1. \tag{7.35}$$

We proceed by induction assuming that $V_j^{\mathrm{T}}r_i = 0$ for $j = 1, \ldots, i-1$. From (7.32) and (7.33) we must have

$$V_j^{\mathrm{T}}r_i - V_j^{\mathrm{T}}AP_id_i = 0, \qquad j = 1, \ldots, i. \tag{7.36}$$

Assuming that $V_i^{\mathrm{T}}AP_i$ is nonsingular, (7.36) implies that, for $j = i$,

$$d_i = (V_i^{\mathrm{T}}AP_i)^{-1}V_i^{\mathrm{T}}r_i. \tag{7.37}$$

In order to satisfy (7.36) also for $j < i$, let us define P_i as follows:

$$P_i = H_i^{\mathrm{T}}Z_i, \tag{7.38}$$

where H_i is an n by n matrix to be defined later and Z_i is an n by m_i matrix, arbitrary save for the condition

$$\det(V_i^{\mathrm{T}}AH_i^{\mathrm{T}}Z_i) \neq 0. \tag{7.39}$$

By induction, (7.36) implies that, for $j < i$,

$$V_j^{\mathrm{T}}AH_i^{\mathrm{T}}Z_id_i = 0. \tag{7.40}$$

For $i = 2$, writing $H_2 = H_1 + D_1$, with H_1 arbitrary nonsingular, (7.40) is satisfied if

$$D_1A^{\mathrm{T}}V_1 = -H_1A^{\mathrm{T}}V_1. \tag{7.41}$$

Choosing an n by m_1 matrix W_1^{T} which is a left inverse of $H_1A^{\mathrm{T}}V_1$, a solution of (7.40) can be given in the form

$$D_1 = -H_1A^{\mathrm{T}}V_1W_1^{\mathrm{T}}H_1. \tag{7.42}$$

For $i > 2$ we can assume by induction that (7.40) is true for the index $i - 1$ $(j < i - 1)$. Writing $H_i = H_{i-1} + D_{i-1}$, (7.40) is satisfied for $j = i - 1$ if

$$D_{i-1}A^{\mathrm{T}}V_{i-1} = -H_{i-1}A^{\mathrm{T}}V_{i-1} \tag{7.43}$$

Again choosing an n by m_{i-1} matrix W_{i-1}^{T} which is a left inverse of $H_{i-1}A^{\mathrm{T}}V_{i-1}$, satisfying

$$W_{i-1}^{\mathrm{T}}H_{i-1}A^{\mathrm{T}}V_{i-1} = I_{m_{i-1}}, \tag{7.44}$$

a solution of (7.43) can be written in the form

$$D_{i-1} = -H_{i-1}A^{\mathrm{T}}V_{i-1}W_{i-1}^{\mathrm{T}}H_{i-1}. \tag{7.45}$$

Thus

$$H_i = H_{i-1} - H_{i-1}A^{\mathrm{T}}V_{i-1}W_{i-1}^{\mathrm{T}}H_{i-1} \tag{7.46}$$

and (7.40) is satisfied for $j < i-1$ by the induction assumption. Assuming that, for $i = 1, \ldots, p$ matrices Z_i, W_i have been given satisfying conditions (7.39) and (7.44), the m-dimensional vector r_{p+1} is orthogonal to the m columns of V and thus x_{p+1} solves the given system.

The next theorem gives a number of properties of the block ABS algorithm. We omit the proofs, since they are similar to the proofs of analogous theorems in Chapter 3.

Theorem 7.10
Let $A \in R^{m,n}$ be full rank; let $V = (V_1, \ldots, V_p) \in R^{m,m}$ and $H_1 \in R^{n,n}$ be nonsingular. For $i = 1, \ldots, p$ define the sequence H_i by

$$H_{i+1} = H_i - H_iA^{\mathrm{T}}V_iW_i^{\mathrm{T}}H_i, \tag{7.47}$$

where $W_i \in R^{n,m_i}$ is a (full-rank) matrix arbitrary save for the condition

$$W_i^{\mathrm{T}}H_iA^{\mathrm{T}}V_i = I_{m_i}. \tag{7.48}$$

Then the recursion (7.47) is well defined, any left inverse of $H_iA^{\mathrm{T}}V_i$ can be taken to satisfy (7.48) and the following relations are true:

$$H_iH_1^{-1}H_j = H_jH_1^{-1}H_i = H_i, \qquad j \leq i, \tag{7.49}$$

$$H_iA^{\mathrm{T}}V_j = 0, \qquad\qquad j < i, \tag{7.50}$$

$$H_i^{\mathrm{T}}W_j = 0, \qquad\qquad j < i, \tag{7.51}$$

$$\mathrm{rank}(H_i) = \mathrm{rank}(H_iA^{\mathrm{T}}V_i, \ldots, H_iA^{\mathrm{T}}V_p) = n - (m_1 + \ldots + m_{i-1}), \tag{7.52}$$

$$\mathrm{rank}(H_iA^{\mathrm{T}}V_j) = m_j, \quad i \leq j \leq p. \tag{7.53}$$

Remark 7.6
If we take $Z_i = W_i S_i$ with $S_i \in R^{i,i}$ arbitrary nonsingular, then condition (7.39) is satisfied, since by (7.48) $\det(W_i^T H_i A^T V_i) = 1$ and thus $\det(Z_i^T H_i A^T V_i) = \det S_i \neq 0$.

It follows from the previous results that the following ALGORITHM 6 is well defined and solves the system $Ax = b$ in no more than p steps (assuming A to be full rank).

ALGORITHM 6: The Block Scaled ABS Algorithm
(A6) Let $x_1 \in R^n$ be arbitrary and $H_1 \in R^{n,n}$ be arbitrary nonsingular. Let $m_1, \ldots, m_p$ be positive integers such that $m_1 + \ldots + m_p = m$; set $i = 1$.
(B6) Compute the residual $r_i = r(x_i)$.
(C6) If $r_i = 0$ stop. x_i solves the system.
(D6) Let $V_i \in R^{m,m_i}$ be an arbitrary matrix, save that $(V_1, \ldots, V_i)$ be full rank. Define the matrix $P_i \in R^{n,m_i}$ by (7.38) where $Z_i \in R^{n,m_i}$ is arbitrary save for the condition (7.39). Compute the vector d_i by solving the system

$$V_i^T A P_i d_i = V_i^T r_i. \tag{7.54}$$

(E6) Update the approximation of the solution by (7.32). If $i = p$, stop, x_{p+1} solves the system.
(F6) Update the matrix H_i by (7.47), with $W_i \in R^{n,m_i}$ arbitrary save for condition (7.48).
(G6) Increment the index i by one and go to (B6).

Remark 7.7
If A is rank deficient, then some of the matrices $A^T V_i$ are rank deficient; thus (7.39) and (7.48) cannot be satisfied.

Remark 7.8
The block scaled ABS algorithm does not represent a generalization of the scaled ABS algorithm. Indeed, let $\bar{x}_1, \ldots, \bar{x}_m$ be the iterates generated by the scaled ABS algorithm with $\bar{x}_1 = x_1$ and let V be the same matrix in both algorithms. By Theorem 7.5, z_1 can be chosen so that $\bar{x}_2$ is an arbitrary point in the linear variety S containing all vectors x such that the residual $r(x)$ is orthogonal to v_1. Since $x_2 \in S$, we can thus have $\bar{x}_2 = x_2$, implying that for the next $m_1 - 1$ steps the residual, and hence the step size, is zero. Thus $\bar{x}_2 = \bar{x}_3 = \ldots = \bar{x}_{m_1+1} = x_2$. By induction the result can be carried on to the following iterations. In the nonlinear case, however, the block formulation provides an essential generalization, as discussed later.

We give now some additional theorems.

Theorem 7.11
The pair (V,P) generated by ALGORITHM 6 is block regular A-semiconjugate.

Proof
For $j < i$ we have by definition of P_j

$$(V_j^T A P_i)^T = Z_i^T (H_i A^T V_j) \tag{7.55}$$

and $H_i A^T V_j$ is zero by (7.50). Since $V_i^T A P_i$ is nonsingular from condition (7.39), the theorem follows. Q.E.D.

The proof of the next theorem is omitted, since it is similar to the proofs of theorems 7.7, 4.2 and 4.3.

Theorem 7.12

For $1 \leqslant i \leqslant p$ let $k = \displaystyle\sum_{j=1}^{i} m_j$ and define the matrices $V^i \in R^{m,k}$ and $W^i \in R^{n,k}$ by

$$V^i = (V_1, \ldots, V_i), \tag{7.56}$$

$$W^i = (W_1, \ldots, W_i). \tag{7.57}$$

Define also the matrix $Q^i \in R^{m_i, m_i}$ by

$$Q^i = W^{iT} H_1 A^T V^i. \tag{7.58}$$

Then the matrix Q^i is strongly nonsingular. Moreover the matrix H_{i+1} has the form

$$H_{i+1} = H_1 - H_1 A^T V^i (Q^i)^{-1} (W^i)^T H_1 \tag{7.59}$$

The following results complement Theorem 7.11 giving a fundamental characterization of the block scaled ABS algorithm.

Theorem 7.13

Let $A \in R^{m,n}$ be full rank; let $V \in R^{m,m}$ be nonsingular and partitioned as in Definition 7.4. Then by suitable choice of the matrix V_i and Z_i in ALGORITHM 6 all possible block regular A-semiconjugate pairs (V, U) with respect to the given partition can be generated.

Proof

Let $U = (U_1, \ldots, U_p)$ be a given matrix such that the pair (V, U) is block A-semiconjugate. Consider the sequence $H_1, \ldots, H_i$ generated by ALGORITHM 6 with $H_1 = I$, $V_1, \ldots, V_i$ the first i submatrices in the partition of V and let $W_1, \ldots, W_{i-1}$ be arbitrary matrices satisfying (7.48). Define W^{i-1}, V^{i-1} and Q^{i-1} as in Theorem 7.12. Then from (7.59) we get

$$H_i^T U_i = U_i - W^{i-1} (Q^{i-1})^{-T} (V^{i-1})^T A U_i \tag{7.60}$$

Now

$$(V^{i-1})^T A U_i = \begin{bmatrix} V_1^T A U_i \\ \cdots \\ V_{i-1}^T A U_i \end{bmatrix}. \tag{7.61}$$

By assumption the matrix on the right-hand side in (7.61) is identically zero. Thus (7.60) yields

$$H_i^T U_i = U_i. \tag{7.62}$$

Now the choice $Z_i = U_i$ is admissible, since $V_i^T A H_i^T Z_i = V_i^T A H_i^T U_i = V_i^T A U_i$ is nonsingular by assumption. With such a choice, $P_i = U_i$, establishing the theorem. Q.E.D.

Corollary 7.1
All possible block regular A-conjugate pairs (V, U) with respect to a given partition can be generated by ALGORITHM 6.

A sufficient condition for ALGORITHM 6 to generate block (regular) A-conjugate pairs (V, P) is given by the following.

Theorem 7.14
A sufficient condition for the pair (V, P) generated by ALGORITHM 6 to be block (regular) A-conjugate is that the matrix Z_i satisfies, for $i = 1, \ldots, p-1$,

$$\text{Range}(Z_i) \subset {}^{\perp}\text{Range}(H_i A^T V_{i+1}, \ldots, H_i A^T V_p) \tag{7.63}$$

Proof
Just apply the definition of block A-conjugacy. Q.E.D.

When the pair (V, P) is block regular A-conjugate, an extension of the Hestenes–Stiefel inversion formula is given by the following theorem.

Theorem 7.15
Let $A \in R^{n,n}$ be nonsingular and let the pair (V, P) be block regular A-conjugate. Then the following identity is true:

$$A^{-1} = \sum_{j=1}^{p} P_j (V_j^T A P_j)^{-1} V_j^T. \tag{7.64}$$

Proof
It is enough to show that we get an identity if both members in (7.64) are postmultiplied by a nonsingular matrix Q. Taking $Q = AP$, we get

$$P = \sum_{j=1}^{p} P_j(V_j^T A P_j)^{-1} V_j^T A P. \tag{7.65}$$

Since the pair (V, P) is block A-conjugate, we have

$$V_j^T A P = (0, \ldots, 0, V_j^T A P_j, 0, \ldots, 0), \tag{7.66}$$

so that (7.65) reads

$$P = \sum_{j=1}^{p} P_j(0, \ldots, 0, I_{m_j}, 0, \ldots, 0) \tag{7.67}$$

which is an identity. Q.E.D.

Finally the following theorem characterizes a block extension of Theorem 5.1.

Theorem 7.16
If H_1 is symmetric positive definite, if the sequence H_i consists of symmetric matrices for $i = 2, \ldots, p$ and if, for $i = 1, \ldots, p$, $Z_i = A^T V_i$, then the pair (P, P) is block H_1^{-1}-conjugate.

Proof
For $j < i$ we have $P_i^T H_1^{-1} P_i = V_j^T A H_j H_1^{-1} H_i A^T V_i = V_j^T A H_i A^T V_i = 0$ by (7.49) and (7.50). A similar result is valid for $i < j$. For $i = j$ we have $P_i^T H_1^{-1} P_i = (H_1^{-1/2} H_i A^T V_i)^T (H_1^{-1/2} H_i A^T V_i)$, which is positive definite since $H_i A^T V_i$ is full rank by (7.53). Q.E.D.

7.6 SEMICONJUGATE AND BICONJUGATE ABS ALGORITHMS: EQUIVALENCE RELATIONS AND VARIATIONAL CHARACTERIZATION

In this section we establish that for any ABS algorithm generating an A-semiconjugate pair (V, P) there exists an ABS algorithm generating an A-conjugate pair $(\overline{V}, \overline{P})$, which is equivalent in the sense that $P = \overline{P}$ and the iterates $\overline{x}_i$, x_i are identical. We also show that for any ABS algorithm the solution x^+ is monotonically and optimally approached from below in a certain weighted Euclidean norm.

We start with the following.

Lemma 7.1
Suppose that the solution $x^+ = x_{m+1}$ of the system $Ax = b$, $A \in R^{m,n}$, is obtained by the iteration $x_{i+1} = x_i - \beta_i p_i$, $i = 1, \ldots, m$, with $x_1 = 0$. Suppose that the vectors p_i are Y-conjugate, Y being symmetric positive definite. Then the following properties are true for $i = 1, \ldots, m$.

(a) The step sizes β_i are uniquely defined.
(b) The solution is monotonically approached from below in the Y-weighted Euclidean norm,

$$\|x_i\| \leqslant \|x_{i+1}\|, \tag{7.68}$$

$$\|x_{i+1} - x^+\| \leqslant \|x_i - x^+\|. \tag{7.69}$$

Proof

We have the identity $x^+ = -\sum_{j=1}^{m} \beta_j p_j$. By premultiplying by Yp_i, we get $\beta_i = -p_i^T Y x^+ / p_i^T Y p_i$, establishing (a). We obtain (7.68) and (7.69) from the identity

$$\|x_{i+1}\| = \|x_i\| + \beta_i^2 p_i^T Y p_i$$

and

$$\|x_i - x^+\| = \|x_{i+1} - x^+\| + \beta_i^2 p_i^T Y p_i.$$

Q.E.D.

Remark 7.9
Note that $\|x_{i+1}\| - \|x_i\| = \|x_i - x^+\| - \|x_{i+1} - x^+\|$. If x_1 is not zero, but is an arbitrary linear combination of the vectors $p_1, \ldots, p_k$, Lemma 7.1 is still valid for $i = k+1, \ldots, m$.

Remark 7.10
If $m = n$, the matrix Y of Lemma 7.1 can be taken as $Y = P^{-T}P^{-1}$ which is positive definite and $P^T Y P = I$, implying that the search vectors are normalized Y-conjugate vectors.

Theorem 7.17
Let $m = n$ and let $Y \in R^{n,n}$ be symmetric positive definite. Consider an ABS method generating search vectors p_i with the scaling vectors v_i given by

$$v_i = \delta_i A^{-T} Y p_i \tag{7.70}$$

with δ_i a nonzero scalar. Let also $x^k \in R^n$ be an arbitrary vector of the form

$$x^k = x_1 - \sum_{i=1}^{k} \beta_i p_i, \tag{7.71}$$

where the β_i are arbitrary scalars, and x_1 is the starting vector of the considered ABS method. Then the following properties are true.
(a) The vectors p_i are Y-conjugate.
(b) The convex quadratic function F_k given by

$$F_k = (x^k - x^+)^T Y (x^k - x^+) \tag{7.72}$$

is minimized with respect to the β_i by the values $\beta_i = \alpha_i$, α_i being the step sizes of the considered ABS method.

Proof
From (7.70) we have $Yp_i = A^T v_i/\delta_i$. For $i < k$, $p_k^T Y p_i = z_k^T H_k A^T v_i/\delta_i$ is zero because of (7.12); for $i > k$, the same follows from the symmetry of Y; for $i = k$, $p_i^T Y p_i > 0$ since $p_i \neq 0$ and $Y > 0$; hence property (a) is true. To prove property (b), we proceed by induction. The derivative of F_k with respect to β_i has the form

$$\frac{\delta F_k}{\delta \beta_i} = -2 p_i^T Y (x^k - x^+). \tag{7.73}$$

For $i = 1$, we can write

$$\frac{\delta F_k}{\delta \beta_1} = -2 p_i^T Y \left(x_1 - x^+ - \beta_1 p_1 - \sum_{j=2}^{k} \beta_j p_j \right). \tag{7.74}$$

Setting the derivative to zero, we get for the minimizing β_1^+

$$\beta_1^+ = \frac{p_1^T Y (x_1 - x^+)}{p_1^T Y p_1} - \sum_{j=2}^{k} \frac{\beta_j p_1^T Y p_j}{p_1^T Y p_1}. \tag{7.75}$$

The second term in (7.75) is zero because of property (a), while the first term, from (7.70) and the identity $x_1 - x^+ = A^{-1} r_1$, takes the form $\beta_1^+ = v_1^T r_1/v_1^T A p_1 = \alpha_1$. Assuming now that $\beta_j = \alpha_j$ for $j = 1, \ldots, i-1$, we can write (7.73) in the form

$$\frac{\delta F_k}{\delta \beta_i} = -2 p_i^T Y \left(x_i - x^+ - \beta_i p_i - \sum_{j=i+1}^{k} \beta_j p_j \right), \tag{7.76}$$

where x_i is the estimate of the solution computed by the considered ABS method at the $(i-1)$th step, and the summation should be omitted if $i = k$. By zeroing the derivative and proceeding as before, we again find that $\beta_i^+ = \alpha_i$. Q.E.D.

Remark 7.11
The function F_k is the Y-weighted distance of x^k for x^+; hence property (b) gives a stronger characterization than the monotonicity result in (7.69). Note also that the special choice for Y implies no assumption on x_1.

Remark 7.12
Theorem 7.17 is still true if we consider an ABS algorithm with the parameter choice $H_1 = I$, $w_i = z_i/z_i^T H_i A^T v_i$ and v_i given by

$$v_i = \delta_i A^{-1} Y z_i \tag{7.77}$$

with δ_i a nonzero scalar. If $m < n$, it is also valid if A^{-T} is replaced by $(A^+)^T$ and we assume the condition $Y p_i \in \text{Range}(A^T)$, if (7.70) holds, or $Y z_i \in \text{Range}(A^T)$, if (7.77) holds. Y can be taken as a positive semidefinite matrix. For proofs see Bodon (1989).

A characterization of the ABS algorithm generating A-conjugate (V, P) pair is given by the following theorem.

Theorem 7.18
Let $m = n$. If the scaling vectors of the ABS method have the form (7.70), with Y symmetric positive definite, then the matrix L in the implicit factorization (7.15) is diagonal. Conversely, if L is diagonal in (7.15), then the scaling vectors can be put in the form (7.70) with Y symmetric positive definite.

Proof
From (7.70) we have $V = A^{-T} Y P \overline{D}$, with $\overline{D}$ diagonal, $\overline{D}_{i,i} = \delta_i$. Thus $V^T A P = \overline{D} P^T Y P$ is symmetric, implying that L in (7.15) is diagonal. If $L = D$ is diagonal, we can assume that $D = I$, after substituting P by $\overline{P} = P D^{-1}$, so that $A^T V = \overline{P}^{-T}$. Let $Y = A^T V \overline{P}^{-1} = \overline{P}^{-T} \overline{P}^{-1} > 0$. Then $V = A^{-T} Y \overline{P} = A^{-T} Y P D^{-1}$ has the required form. Q.E.D.

The following theorems give conditions for different scaling vectors to generate the same search vectors and the same approximations to the solution, say to generate equivalent algorithms (in exact arithmetic).

Theorem 7.19
Consider the sequences p_i, $\overline{p}_i$ generated by two ABS algorithms with the parameter choices (H_1, z_i, w_i, v_i) and $(H_1, z_i, \overline{w}_i, \overline{v}_i)$, with $\overline{v}_i$ given by

$$\overline{v}_i = \sum_{j=1}^{i} U_{i,j} v_j, \quad U_{i,i} = 1, \tag{7.78}$$

or

$$\overline{V} = VU, \tag{7.79}$$

with the U unit upper triangular. Then, if w_i satisfies (7.10), $\overline{w}_i = w_i$ also satisfies (7.10) and with such a choice the sequences p_i and $\overline{p}_i$ are identical.

Proof
Since $H_1 = \overline{H}_1$ the theorem is obvious for $i = 1$. Assume by induction that $H_j = \overline{H}_j$ up to $j = i$. Then $\overline{H}_iA^T\overline{V}_i = H_iA^Tv_i + \sum_{j=1}^{i-1} U_{i,j}H_iA^Tv_j = H_iA^TV_i$ from (7.12). It follows immediately that $\overline{w}_i^T\overline{H}_iA^Tv_i = w_i^TH_iA^Tv_i$ and that $\overline{H}_{i+1} = H_{i+1}$, implying that $\overline{p}_{i+1} = p_{i+1}$. Q.E.D.

Remark 7.13
Theorem 7.19 can be proved under even less restrictive conditions on the parameters. For instance we can assume, without loss of generality, that w_i and z_i are multiples of each other. Then the theorem remains true if $\overline{w}_i$ and $\overline{z}_i$ are multiples of each other and $\overline{z}_i = \sum_{j=1}^{i} U'_{i,j}z_j$, $U'_{i,i} = 1$.

The converse of the result given in Theorem 7.19 is stated by the following theorem.

Theorem 7.20
If two ABS algorithms generate the same sets of search vectors $P = \overline{P}$, then their scaling vectors V, $\overline{V}$ satisfy (7.79) with U an upper triangular matrix.

Proof
From the assumption of the theorem (7.15) reads respectively $V^TAP = L$, $\overline{V}^TAP = \overline{L}$. From the first relation we have $AP = V^{-T}L$. Substituting in the second relation, we have $\overline{V} = VL^{-T}\overline{L}^T$, proving the theorem with $U = L^{-T}\overline{L}^T$. Q.E.D.

The following theorem establishes the equivalence of a general scaled ABS algorithm with some ABS algorithm generating an A-conjugate pair (V, P).

Theorem 7.21
Consider an ABS algorithm generating an A-semiconjugate pair (V, P). Then there exists an ABS algorithm generating an A-conjugate pair $(\overline{V}, P)$.

Proof
For the given algorithm generating the A-semiconjugate pair (V, P), (7.15) can be written in the form $V^TAP = \overline{L}D$, with $\overline{L}$ unit lower triangular, D diagonal, $D_{i,i} = 1/L_{i,i}$. Consider the ABS algorithm which has the same parameters as the given algorithm, save for the scaling matrix, which is given by $\overline{V} = VL^{-T}$. From Theorem

7.19 this algorithm generates the same search vectors; hence (7.15) reads $\overline{V}^T AP = \overline{L}$. However, $\overline{V}^T AP = \overline{L}^{-1} V^T AP = \overline{L}^{-1} \overline{L} D = D$; hence $(\overline{V}, P)$ is an A-conjugate pair. Q.E.D.

Corollary 7.2
Every unscaled ABS algorithm generates the same search vectors produced by a scaled ABS algorithm where V is upper triangular and the pair (V, P) is A-conjugate.

Theorem 7.22
Consider two ABS algorithms where the set of search and scaling vectors satisfy $\overline{P} = P$, $\overline{V} = VU$, with U nonsingular upper triangular. Then, if $\overline{x}_1 = x_1$, the identity $\overline{x}_i = x_i$ is valid for all indexes until termination.

Proof
It is enough to show the identity of the step sizes. For $i = 1$ this is immediate.

Assuming that $\overline{\alpha}_j = \alpha_j$ for $j < i$, we have $\overline{r}_i = r_i$; hence $\overline{\alpha}_i = r_i^T \overline{v}_i / p_i^T A \overline{v}_i = \left(U_{i,i} r_i^T v_i + \right.$

$$\sum_{j=1}^{i-1} U_{i,j} r_i^T A^T v_j \Bigg). \text{ However, } r_i^T v_j = 0 \text{ from Theorem 7.3 and } p_i^T A^T v_j = z_i^T H_i A^T v_j = 0$$

from (7.12), implying that $\overline{\alpha}_i = r_i^T v_i / p_i^T A^T v_i = \alpha_i$. Q.E.D.

7.7   RELATIONSHIPS WITH THE STEWART GENERALIZED CONJUGATE DIRECTION ALGORITHM

In this section we introduce the generalized conjugate direction algorithm for linear systems of Stewart (1973). We analyse the relationships with the scaled ABS algorithm, showing that the two algorithms are equivalent, in the sense of generating the same sets of search vectors and approximations of the solution. The Stewart algorithm differs in algebraic formulation from the ABS algorithm, being based upon an extension of the Fox *et al.* (1948) conjugation procedure.

Stewart has introduced two classes of algorithms for solving the system $Ax = b$, $A \in R^{m,n}$ and full rank, $x \in R^n$, $b \in R^m$, $m \leq n$. These are the following.

ALGORITHM 7: The Stewart Algorithm I
(A7) Let $x_1 \in R^n$ be arbitrary. Let $V = (v_1, \ldots, v_m) \in R^{m,m}$ be nonsingular, let $U = (u_1, \ldots, u_m) \in R^{n,m}$ be full rank and assume that the pair (V, U) is regular A-conjugate. Set $i = 1$.
(B7) Compute the residual $r_i = r(x_i)$.
(C7) If $r_i = 0$, stop, x_i solves the system.
(D7) Update the estimate of the solution by

$$x_{i+1} = x_i - \frac{v_i^T r_i}{v_i^T A u_i} u_i. \tag{7.80}$$

(E7) Increment the index i by one and go to (B7).

ALGORITHM 8: The Stewart Algorithm II
(A8) Let $x_1 \in R^n$ be arbitrary. Let $V = (V_1, \ldots, V_p) \in R^{m,m}$ be nonsingular, with $V_i \in R^{m,m_i}$, let $U = (U_1, \ldots, U_p) \in R^{n,m}$ be full rank, with $U_i \in R^{n,m_i}$ and assume that the pair (V, U) is block regular A-semiconjugate with respect to the given partition. Set $i = 1$.
(B8) Compute the residual $r_i = r(x_i)$.
(C8) If $r_i = 0$, stop, x_i solves the system.
(D8) Compute the vector $d_i \in R^{m_i}$ by solving the system

$$(V_i^T A U_i) d_i = V_i^T r_i \tag{7.81}$$

(E8) Update the approximation of the solution by

$$x_{i+1} = x_i - U_i d_i. \tag{7.82}$$

(G8) Increment the index i by one and go to (B8).

In the Stewart algorithms the pairs (V, U) are assumed to be given. ALGORITHM 7 is a special case of ALGORITHM 8 as the scaled ABS algorithm is a special case of the block scaled ABS algorithm. The block scaled ABS algorithm is a particular case of ALGORITHM 8, since, letting $U = P$, the pair (V, U) is block regular A-semiconjugate. However, from Theorem 7.15 every block regular A-semiconjugate pair (V, U) can be generated by the block scaled ABS algorithms; hence the two classes of algorithms are equivalent.

The construction of the matrix U in ALGORITHM 8 is not considered by Stewart, while the following procedure is given for ALGORITHM 7. Let A and V be given and let $D = (d_1, \ldots, d_m) \in R^{n,m}$ be an arbitrary full rank matrix. The idea, as in the Gram–Schmidt procedure, is to construct u_i as a linear combination of $d_1, \ldots, d_i$. This process is called by Stewart the A-conjugation of D with respect to V. The conditions which guarantee that it can be carried on are given by the following theorem.

Theorem 7.23
Let $V \in R^{m,m}$, $A \in R^{m,n}$ and $R \in R^{n,m}$ be full rank. Then the necessary and sufficient conditions for the A-conjugation of D with respect to V to be possible is that $V^T A D$ be strongly nonsingular.

Proof
For necessity, observe that the A-conjugation of D requires finding an upper triangular matrix S such that

$$U = DS. \tag{7.83}$$

Since the pair (V, U) is regular A-semiconjugate, it follows that the matrix

$L = V^\mathrm{T}AU = V^\mathrm{T}ADS$ is nonsingular lower triangular. Then S must also be nonsingular, so that $V^\mathrm{T}AD = LS^{-1}$. Since $R = S^{-1}$ is upper triangular, $V^\mathrm{T}AD$ is strongly nonsingular, being LR decomposable. For sufficiency, just observe that, if $V^\mathrm{T}AD = LR$, then the pair (V, U), where $U = DR^{-1}$, is regular A-semiconjugate. Q.E.D.

Remark 7.14
From a theorem of Householder (1964) the matrix S in (7.83) is uniquely determined up to the scaling of the rows. Thus the columns of U, and hence the search vectors u_i in ALGORITHM 7, have a uniquely defined direction, for a given D.

Remark 7.15
By setting $D = US^{-1}$, S nonsingular upper triangular, any regular A-semiconjugate pair (V, U) can be generated by the Stewart A-conjugation process. Thus this process is equivalent, in the sense of generating search vectors with the same direction, to the scaled ABS algorithm.

An implementation of the Stewart A-conjugation process is given by the following.

ALGORITHM 9: The Stewart A-conjugation Process

(A9) Let $V = (v_1, \ldots, v_m) \in R^{m,m}$, $D = (d_1, \ldots, d_m) \in R^{n,m}$, $1 < m \leqslant n$, be such that $V^\mathrm{T}AD$ is strongly nonsingular. Compute $u_1 \in R^n$ by $u_1 = \delta_1 d_1$, δ_1 a nonzero scalar. Define the triangular matrix $L^1 \in R^{1,1}$ by $L^1 = v_1^\mathrm{T}Au$. Set $i = 2$.

(B9) If $i = m + 1$, stop; otherwise compute $s_{i-1} \in R^{i-1}$ by solving the linear (triangular) system

$$L^{i-1}s_{i-1} = (V^{i-1})^\mathrm{T}Ad_i \tag{7.84}$$

where $V^{i-1} \in R^{m,i-1}$ consists of the first $i - 1$ columns of V.

(C9) Compute $u_i \in R^n$ by

$$u_i = \delta_i(d_i - U^{i-1}s_{i-1}) \tag{7.85}$$

where δ_i is an arbitrary nonzero scalar and $U^{i-1} \in R^{n,i-1}$ is the matrix whose columns are $u_1, \ldots, u_{i-1}$.

(D9) Compute the triangular matrix $L^i \in R^{i,i}$ by

$$L^i = (V^i)^\mathrm{T}AU^i. \tag{7.86}$$

(E9) Increment the index i by one and go to (B9).

Remark 7.16
The number of multiplications required by ALGORITHM 9 is of the order of $(11/6)m^3 + (3/2)m^2n$, say, for $m = n$, $(10/3)n^3$. The corresponding order in the scaled ABS algorithm is (for $m = n$) $4n^3$ for arbitrary choices of z_i and w_i, $3n^3$ if, without loss of generality, z_i is a multiple of w_i.

Remark 7.17

Checking the strong nonsingularity of the matrix $V^T AD$ is not as simple as satisfying conditions (7.6) and (7.10).

Remark 7.18

In the ith step of ALGORITHM 9 the vector u_i is built using only the first $i - 1$ columns of V, the ith column being used in the last step for the update of L^i. Thus, as in the scaled ABS algorithm, v_i can be interpreted as a parameter available at the ith iteration, free save that $v_1, \ldots, v_i$ must be linearly independent. Also note that v_i can be a function of $u_1, \ldots, u_i$.

Remark 7.19

If in ALGORITHM 9 we take $D = (e_1, \ldots, e_m)$ and $V = (u_1, \ldots, u_m)$, then the vectors u_i coincide in direction with the vectors generated by the conjugation method of Fox *et al.* (1948), also considered by Hestenes and Stiefel (1952). The search vectors p_i generated by the implicit LU–LL^T algorithm described in Chapter 6 are regular A-semiconjugate and p_i, being equal to the ith row of the matrix H_i, is a linear combination of $e_1, \ldots, e_i$. From Remark 7.9 it follows that they have the same direction as the vectors generated by the Fox *et al.* algorithm. Thus the Fox *et al.* algorithm is also equivalent, in the sense of generating the same search directions, to the escalator method (as was already observed from numerical experiments by Fox *et al.*), the linear Brown method and the Sloboda method.

7.8   RELATIONSHIPS WITH THE BROYDEN GENERAL FINITELY TERMINATING ALGORITHM

A class of algorithms with finite termination for solving the linear system $Ax = b$, x, $b \in R^n$, A nonsingular, has been studied by Broyden (1985), in connection with the problem of determining optimally stable algorithms in the sense defined by Broyden (1974). The Broyden class is based upon the following procedure.

ALGORITHM 10: The Broyden General Finitely Terminating Algorithm

(A10) Let $x_1 \in R^n$ be arbitrary and set $i = 1$.

(B10) Compute the residual $r_i = r(x_i)$. If $r_i = 0$, stop, x_i solves the system.

(C10) Let $v_i, p_i \in R^n$ satisfy condition $v_i^T A p_i \neq 0$. Compute the scalar α_i by

$$\alpha_i = \frac{v_i^T r_i}{v_i^T A p_i.} \tag{7.87}$$

(D10) Update the estimate of the solution by

$$x_{i+1} = x_i - \alpha_i p_i. \tag{7.88}$$

(E10) Increment the index i by one and go to (B10).

The conditions that the vectors p_i and v_i must satisfy in order that x_{n+1} solves the system for any starting point x_1 are given in the following theorem of Broyden (1985), as amended by Noble and Galantai.

Theorem 7.23

Suppose that n steps are performed by ALGORITHM 10 where the nonsingular matrices $V = (v_1, \ldots, v_n)$, $P = (p_1, \ldots, p_n)$ are such that $v_i^{\mathrm{T}} A p_i \# 0$ for $i = 1, \ldots, n$. Then the necessary and sufficient condition for the vector x_{n+1} to solve $Ax = b$ for any starting point x_1 is that the matrix $L = V^{\mathrm{T}} A P$ is nonsingular lower triangular.

Proof

First we prove necessity. Define the matrices $\overline{P}_i$, $\overline{V}_i \in R^{n, n-i+1}$ by

$$\overline{P}_i = (p_i, \ldots, p_n), \tag{7.89}$$

$$\overline{V}_i = (v_i, \ldots, v_n). \tag{7.90}$$

Define the vector $\overline{a}_i \in R^{n-i+1}$ by

$$\overline{a}_i = (\alpha_i, \ldots, \alpha_n). \tag{7.91}$$

Now for $i \leqslant j \leqslant n$ the vector x_{j+1} computed by ALGORITHM 10 has the form

$$x_{j+1} = x_i - \sum_{k=i}^{j} \alpha_k p_k \tag{7.92}$$

and the residual r_{j+1} is

$$r_{j+1} = r_i - A \sum_{k=i}^{j} \alpha_k p_k. \tag{7.93}$$

By definition of α_i the residual r_{i+1} is orthogonal to v_i. Thus premultiplying (7.93) by v_j yields

$$v_j^{\mathrm{T}} r_i + v_j^{\mathrm{T}} A \sum_{k=i}^{j} \alpha_k p_k = 0. \tag{7.94}$$

Let $\overline{L}_i$ be the lower triangular matrix with elements

$$(\overline{L}_i)_{j,k} = v_{i+j-1}^{\mathrm{T}} A p_{i+k-1}, \ k \leqslant j, \ j = 1, \ldots, n-i+1,$$

and note that by assumption $\overline{L}_i$ is nonsingular. Then, if we write (7.94) for $j = i, i+1, \ldots, n$, the obtained $n-i+1$ equations can be put in the form

$$\bar{L}_i\bar{a}_i = -V_i^{\mathrm{T}}r_i, \tag{7.95}$$

implying that

$$\bar{a}_i = -(\bar{L}_i)^{-1}V_i^{\mathrm{T}}r_i. \tag{7.96}$$

Then, using (7.96), (7.92) can be written as

$$x_{j+1} = x_i - \bar{P}_i(\bar{L}_i)^{-1}V_i^{\mathrm{T}}r_i \tag{7.97}$$

Writing (7.97) for $j = n$ and using

$$r_i = A(x_i - x^+), \tag{7.98}$$

where x^+ solves $Ax = b$, we also get

$$x_{n+1} = (I - Q_i)x_i + Q_i x^+, \tag{7.99}$$

where Q_i is given by

$$Q_i = \bar{P}_i(\bar{L}_i)^{-1}V_i^{\mathrm{T}}A. \tag{7.100}$$

Let $i = 1$ and assume that $x_{n+1} = x^+$ holds for arbitrary x_1. Then (7.97) becomes

$$x^+ = x_1 - \bar{P}_1(\bar{L}_1)^{-1}V_1^{\mathrm{T}}A(x_1 - x^+) \tag{7.101}$$

or, letting $z = x^+ - x_1$,

$$z = \bar{P}_1(\bar{L}_1)^{-1}V_1^{\mathrm{T}}Az. \tag{7.102}$$

Since z is arbitrary, it follows that

$$I = \bar{P}_1(\bar{L}_1)^{-1}V_1^{\mathrm{T}}A. \tag{7.103}$$

Postmultiplying by P_1 and premultiplying by P_1^{-1}, we have

$$I = (\bar{L}_1)^{-1}V_1^{\mathrm{T}}A\bar{P}_1. \tag{7.104}$$

Premultiplying by $\bar{L}_1$, we finally have

$$\overline{L}_1 = \overline{V}_1^{\mathrm{T}} A \overline{P}_1, \tag{7.105}$$

where $\overline{L}_1$ is lower triangular. Since $\overline{V}_1 = V$ and $\overline{P}_1 = P$, then the necessity is proved.

For sufficiency, just observe that with the given conditions ALGORITHM 10, with $P = U$, coincides with ALGORITHM 7 and therefore has n-step termination. Q.E.D.

Remark 7.24
Theorem 7.23 can be extended to systems of $m < n$ equations under the assumption that the vectors $Ap_1, \ldots, Ap_m$ are linearly independent.

Remark 7.25
It follows from Theorem 7.23 that, if in ALGORITHM 10 the matrix $V^{\mathrm{T}}AP$ has some nonzero upper diagonal elements, then either more than n iterations are required for termination or termination does not occur for some starting points. Algorithms with finite termination in more than n steps have been proposed in the literature. For instance, it was shown by Gerber and Luk (1981) that the quasi-Newton algorithm of Broyden (1965) for nonlinear systems terminates in no more than $2n$ iterations on linear systems (and there exist systems where $2n$ iterations are actually required). This algorithm differs, *inter alia*, from ALGORITHM 10 in the fact that the step size has the fixed value of one.

7.9 BIBLIOGRAPHICAL REMARKS

The scaled ABS algorithm has been firstly introduced by Abaffy (1984). The derivation in terms of applying the basic ABS algorithm to a scaled system is due to Spedicato and has appeared in the paper by Abaffy and Spedicato (1985). The block ABS algorithm is due to Abaffy and Galantai (1986). The definition of A-semiconjugacy given here is different from that given by Stewart (1973) and by Abaffy and Galantai (1986) and closer to that given by Hegedüs and Bodocs (1982). The relation with the Stewart algorithm was first studied by Abaffy and Galantai (1986). Equations (7.26) and (7.27) and Theorems 7.17, 7.19 and 7.21 are due to Bodon (1989).

8

Subclasses in the scaled ABS algorithm

8.1 INTRODUCTION

In this chapter we consider a number of choices of the scaling matrix V, each one defining a subclass of the scaled ABS algorithm (possibly intersecting with other subclasses). Eight subclasses are considered, indicated by symbols S1, ..., S8. General properties of the subclasses are investigated and various particular algorithms are analysed, among them ABS formulations of classical algorithms such as the QR method and several conjugate gradient type methods.

Subclass S1 consists of algorithms where H_i is symmetric. The Huang algorithm investigated in Chapter 5 is a member of S1 and we also show that the conjugate gradient-type algorithm of Fridman belongs to S1.

Subclasses S2–S6 are characterized by the property that the matrix $L^i = (V^i)^T A P^i$ (see (7.14)), is diagonal, implying that the vectors v_i, p_i form an A-conjugate pair. For subclasses S2–S4 the matrix V^i has the form $V^i = SP^i$ ($S = I$ in S2, $S = A$ in S3 and $S = A^{-T}$ in S4). Well-definiteness requirements are for S2 that A be symmetric positive definite and for S3 and S4 that A be square nonsingular. Well-known classical algorithms are members of these subclasses: the Hestenes–Stiefel and the Lanczos methods belong to S2, the QR method to S3, and the conjugate gradient-type algorithm of Craig to S4. Subclass S4 is moreover shown to be equivalent to subclass S1; hence the Huang and the Fridman algorithms can also be considered as members of S4. The algorithms in S3 reduce at each iteration the Euclidean norm of the residual, those in S4 reduce the Euclidean norm of the error $s_i = x_i - x^+$ and moreover satisfy the Broyden condition for optimal stability.

Subclass S5 consists of algorithms that are realizations of the Voyevodin class of conjugate gradient-type algorithms, where the search vector can be expressed as a linear combination of only two vectors. The Hestenes–Stiefel and the Craig algorithms also belong to S5. Subclass S6 contains algorithms which are realizations of the Hegedüs–Bodocs biorthogonalization procedure, which includes also algorithms of subclass S5.

The algorithms of subclass S7 have the property that the matrix L^i in (7.14) is lower bidiagonal. The subclass contains an algorithm which can be used to reduce A into tridiagonal form by an orthogonal similarity transformation.

Finally the algorithms of subclass S8 are related to a transformation of A into upper Hessenberg form. The algorithms in the last two subclasses may be useful in the context of the eigenvalue–eigenvector problem, an issue which, however, will not be developed in this monograph.

8.2   SUBCLASS S1: THE SCALED SYMMETRIC AND THE SCALED HUANG ALGORITHMS

The algorithms in subclass S1 are characterized by the property that H_i is symmetric. The conditions for H_i to be symmetric are given by the following theorem.

Theorem 8.1

Let A be square nonsingular and let H_i be symmetric. Then H_{i+1} is symmetric if and only if w_i has the form

$$w_i = \sum_{j=1}^{i} \tau_j A^T v_j \, , \tag{8.1}$$

with the τ_j arbitrary scalars.

Proof

H_{i+1} given by (7.9) is symmetric if and only if $H_i^T w_i$ is a multiple of $H_i A^T v_i$. Writing w_i in the form $w_i = \sum_{j=1}^{n} \tau_j A^T v_j$, Theorem 7.2 implies that $H_i w_i = H_i \sum_{j=i}^{n} \tau_j A^T v_j$. By Theorem 7.1 the vectors $H_i A^T v_j$ are linearly independent for $j = i, \ldots, n$; hence the proportionality of $H_i^T w_i$ to $H_i A^T v_i$ holds if and only if $\tau_j = 0$ for $j = i+1, \ldots, n$. Q.E.D.

If $\tau_j = 0$ for $j = 1, \ldots, i-1$, τ_i is determined by condition (7.10), yielding

$$w_i = \frac{A^T v_i}{v_i^T A H_i A^T v_i} \, . \tag{8.2}$$

If $H_1 = I$, using idempotency of H_i we can write (8.2) in the following equivalent form, which generalizes the modified form of the Huang update:

$$w_i = \frac{H_i A^T v_i}{\|H_i A^T v_i\|_2^2} \, . \tag{8.3}$$

If H_1 is positive definite, (8.2) is well defined and the denominator in the right-hand side is positive.

Definition 8.1
The subclass of the scaled ABS algorithm where H_1 is symmetric positive definite and w_i is given by (8.2) is called the class of the scaled symmetric updates.

Definition 8.2
The subclass of the scaled ABS algorithm where $H_1 = I$, w_i is given by (8.2) and z_i is a multiple of w_i is called the class of the scaled Huang algorithms.

Definition 8.3
The subclass of the scaled ABS algorithm where $H_1 = I$, w_i is given by (8.3) and z_i is a multiple of w_i is called the class of the scaled modified Huang algorithms.

   A basic property of the search vectors in Subclass 1 is given by the following reformulation of Theorem 5.1.

Theorem 8.2
If z_i is a multiple of w_i, then the search vectors in subclass S1 are H_1^{-1}-*conjugate*.

Corollary 8.1
The search vectors generated by the scaled Huang or the modified Huang algorithms are orthogonal.

   A fundamental property of the Huang algorithm is that, under certain conditions, the vector x_{i+1} is the vector of the minimal Euclidean norm that solves the first i equations. In the framework of the scaled Huang algorithm this property is reformulated as follows.

Theorem 8.3
Consider the class of the scaled Huang algorithms where x_1 is given by

$$x_1 = \sum_{j=1}^{k} \tau_j A^{\mathrm{T}} v_j \ , \tag{8.4}$$

where the τ_j are scalars and $k < m$. Then for $i \geqslant k$ the vector x_{i+1} has the minimal Euclidean norm among all vectors $\bar{x}$ such that

$$v_j^{\mathrm{T}}(A\bar{x} - b) = 0, \quad j \leqslant i \ . \tag{8.5}$$

   Theorem 5.5 and Corollary 5.2 can be reformulated as follows.

Theorem 8.4
Consider the class of the scaled Huang algorithm where x_1 is the zero vector. Then the sequence $\|x_i\|$ approaches monotonically $\|x^+\|$ from below and the distance $\|x_i - x^+\|$ decreases monotonically.

Remark 8.1

In the class of the scaled Huang algorithms or the modified Huang algorithms, update (7.9) can be written in the form

$$H_{i+1} = H_i - \frac{p_i p_i^{\mathrm{T}}}{p_i^{\mathrm{T}} p_i} . \tag{8.6}$$

We now show that subclass S1 contains the so-called orthogonal direction (OD) method for symmetric indefinite systems attributed to Fridman (1963) and further developed by Fletcher (1975), Paige and Saunders (1975), Stoer and Freund (1982) and Freund (1983).

Given A square nonsingular and symmetric the OD method is defined by the following procedure.

ALGORITHM 11:   The Fridman Orthogonal Direction Method

(A11) Let $\bar{x}_1 \in R^n$ be arbitrary. Compute the residual $\bar{r}_1 = A\bar{x}_1 - b$. Stop if $\bar{r}_1 = 0$; otherwise set $\bar{p}_0 = \bar{r}_1$, $\bar{p}_1 = A\bar{r}_1$ and $i = 1$.

(B11) Update the estimate $\bar{x}_i$ of the solution by

$$\bar{x}_{i+1} = \bar{x}_i - \frac{\bar{r}_i^{\mathrm{T}} \bar{p}_{i-1}}{\bar{p}_i^{\mathrm{T}} \bar{p}_i} \bar{p}_i . \tag{8.7}$$

(C11) Compute the residual $\bar{r}_{i+1}$. If $\bar{r}_{i+1} = 0$, stop.

(D11) Compute the search vector $\bar{p}_{i+1}$ by

$$\bar{p}_{i+1} = A\bar{p}_i - \tau_i \bar{p}_i - \delta_i \bar{p}_{i-1} , \tag{8.8}$$

where

$$\tau_i = \frac{\bar{p}_i^{\mathrm{T}} A \bar{p}_i}{p_i^{\mathrm{T}} \bar{p}_i} , \tag{8.9}$$

$$\delta_i = \frac{\bar{p}_i^{\mathrm{T}} \bar{p}_i}{\bar{p}_{i-1}^{\mathrm{T}} \bar{p}_{i-1}} , \qquad i > 1, \delta_1 = 0. \tag{8.10}$$

(E11) Increment i by one and go to (B11).

Among the properties of the OD method we recall the following, (see Freund (1983) for proofs).

— The algorithm is well defined.
— The search vectors are orthogonal.
— The residual $\bar{r}_i$ is a linear combination of $\bar{p}_0, \ldots, \bar{p}_i$.
— The vectors $\bar{p}_i, \bar{r}_i$ are A^{-1}-semiconjugate.
— The Euclidean norm of the error $s_i = \bar{x}_i - x^+$ decreases monotonically (not necessarily strictly monotonically, since $\bar{x}_{i+1} = \bar{x}_i$ is possible).

— The procedure terminates after a number of iterations equal to the number of eigenvectors corresponding to distinct eigenvalues appearing in the expansion of s_1 in terms of the eigenvectors of A.

In order to prove that a certain algorithm in the class of the scaled Huang algorithms is equivalent to the OD method (in the sense that, given $x_1 = \bar{x}_1$, the sequences x_i and $\bar{x}_i$ are identical), we use the following.

Lemma 8.1
Suppose that A is symmetric and that in the scaled Huang algorithm v_i can be chosen so that

$$p_{i+1} = Ap_i + \sum_{k=1}^{i} \tau_k p_k \tag{8.11}$$

for some scalars τ_k. Then the following identity is true for $j > i$:

$$H_i Ap_j = Ap_j . \tag{8.12}$$

Proof
For $i = 1$ the lemma is trivial. Proceeding by induction, assume that $H_i Ap_j = Ap_j$ for $j > i$. To prove that $H_{i+1} Apj = Ap_j$ for $j > i+1$, observe that, from (8.6) and symmetry,

$$H_{i+1} Ap_j = H_i Ap_j - \frac{(Ap_i)^T p_j}{p_i^T p_i} p_i . \tag{8.13}$$

The first term on the right-hand side equals Ap_j by the induction. Using the orthogonality of the search vectors, and (8.11) to express Ap_i, the second term on the right-hand side is seen to be zero, completing the induction. Q.E.D.

Theorem 8.5
Let A be nonsingular and symmetric and consider the algorithm in the class of the scaled Huang algorithms where v_i is given by the rule $v_1 = r_1$, $v_i = p_{i-1}$ for $i > 1$. Then, if $x_1 = \bar{x}_1$ the sequences p_i, $\bar{p}_i$ and x_i, $\bar{x}_i$ generated by this algorithm and by the OD method are identical.

Proof
We proceed by induction, proving first that $x_2 = \bar{x}_2$, $p_2 = \bar{p}_2$. We have $p_1 = Ar_1 = \bar{p}_1$ and $x_2 = x_1 + (r_1^T v_1/v_1^T Ap_1)p_1 = x_1 + (r_1^T r_1/r_1^T Ap_1)p_1 = x_1 + (r_1^T r_1/p_1^T p_1)p_1 = \bar{x}_2$ and, from (8.4), $p_2 = H_2 Av_2 = (I - p_1 p_1^T/p_1^T p_1)Ap_1 = \bar{p}_2$. In a similar way we can prove the identity of x_3, p_3 with $\bar{x}_3, \bar{p}_3$. Assuming now that $x_j = \bar{x}_j$ and $p_j = \bar{p}_j$ up to the index $j = i$, we prove that $x_{i+1} = \bar{x}_{i+1}$ and $p_{i+1} = \bar{p}_{i+1}$. We first observe that $x_{i+1} = x_i + (r_i^T v_i/v_i^T Ap_i)p_i = \bar{x}_i + (\bar{r}_i^T \bar{p}_{i-1}/\bar{p}_{i-1}^T Ap_i)\bar{p}_i = \bar{x}_{i+1}$, since (8.8) implies that

$$A\bar{p}_{i-1} = \bar{p}_i + \tau_{i-1}\bar{p}_{i-1} + \delta_{i-1}\bar{p}_{i-2}$$

and the vectors $\bar{p}_i$ are orthogonal. Since the induction implies also that

$$Ap_{i-1} = p_i + \tau_{i-1}p_{i-1} + \delta_{i-1}p_{i-1} , \qquad (8.14)$$

we are allowed to use Lemma 8.1. By successive use of the update formula (8.6) we obtain

$$p_{i+1} = H_{i-1}Ap_i - \frac{p_{i-1}{}^T Ap_i}{p_{i-1}{}^T p_{i-1}} p_{i-1} - \frac{p_i{}^T Ap_i}{p_i{}^T p_i} p_i .$$

From Lemma 8.1, (8.14) and the orthogonality of the p_j we obtain

$$p_{i+1} = Ap_i - \frac{p_i{}^T Ap_i}{p_i{}^T p_i} p_i - \frac{p_i{}^T p_i}{p_{i-1}{}^T p_{i-1}} p_{i-1}$$

and the identity $p_{i+1} = \bar{p}_{i+1}$ follows from the induction assumption. Q.E.D.

The OD method is known to be numerically unstable owing to rapid loss of orthogonality of the search vectors. A numerically stable version is that developed by Freund (1983), which is called the stabilized orthogonal direction (STOD) method. In the STOD method the definitions (8.7), (8.9) and (8.10) are changed to

$$x_{i+1} = x_i - \frac{r_i{}^T p_i}{p_i{}^T A^2 p_i} p_i \qquad (8.15)$$

$$\tau_i = \frac{p_i{}^T A^3 p_i}{p_i{}^T A^2 p_i} \qquad (8.16)$$

and

$$\delta_i = \frac{p_i{}^T A^2 p_i}{p_{i-1}{}^T A^2 p_{i-1}} . \qquad (8.17)$$

It can be shown (Abaffy 1988d), that the STOD method is obtained if the vector p_i in the ABS version of the OD method is premultiplied by the matrix H_i. Hence the better numerical stability of the STOD method appears to be related to the reprojection of the search vector on the range of H_i, whose usefulness was already observed for the modified Huang algorithm. Note further that in the ABS formulation of the STOD method there are no extra multiplications with the matrix A, which can be an advantage versus the original STOD method. The multiplications required by the ABS formulation are also fewer.

8.3 SUBCLASS S2: THE CLASS OF THE CONJUGATE DIRECTION ABS ALGORITHM

Let us assume that A is symmetric positive definite. Subclass S2 is obtained by the choice

$$v_i = p_i \ . \tag{8.18}$$

The well-definiteness condition (7.6) is satisfied, since it reads $v_i^T A p_i = p_i^T A p_i > 0$. Choice (8.18) corresponds to taking $Y = A$ and $\delta_i = 1$ in (7.70). We can therefore restate the results in Theorem (7.17) as the following.

Theorem 8.6
Let A be symmetric and positive definite. Then the scaled ABS class where $v_i = p_i$ generates A-conjugate search vectors and the iterate x_{i+1} minimizes over the linear variety $x_1 + \mathrm{Span}(p_1, \ldots, p_i)$ the convex quadratic function

$$F(x) = (x - x^+)^T A (x - x^+) \ . \tag{8.19}$$

Remark 8.2
The positive definiteness of A is a sufficient but not a necessary condition for the well-definiteness of the subclass. In section 8.6 we show indeed the existence of a well-defined algorithm in S2 (the method of complete Hermitian decomposition) for which positive definiteness of A is not required.

Definition 8.4
Subclass S2 is called the ABS class of conjugate direction algorithms.

In subclass S2, z_i and w_i are essentially arbitrary. Symmetric matrices H_i are obtained with H_1 symmetric and w_i given by

$$w_i = \frac{A p_i}{p_i^T A H_i A p_i} \ . \tag{8.20}$$

Choice (8.20) is well defined for $H_1 > 0$. If moreover $H_1 = I$, we can write (8.20) in the form

$$w_i = \frac{A p_i}{\|H_i A p_i\|_2^2} \ . \tag{8.21}$$

Remark 8.3
In order to generate search vectors that are both orthogonal and A-conjugate, we should select z_i a multiple of w_i, i.e.

$$z_i = \frac{Ap_i}{\tau_i} \, , \tag{8.22}$$

where τ_i is an arbitrary nonzero scalar. However, (8.22) is equivalent to

$$AH_i^T z_i = \tau_i z_i \, . \tag{8.23}$$

Hence, in general, z_i cannot be given in closed simple form since it is an eigenvector of AH_i^T. Indeed the requirement on the search vectors of being simultaneously orthogonal $(P^T P = I)$ and A-conjugate $(P^T A P = D)$ implies that they are eigenvectors of A $(AP = PD)$.

In several algorithms that we shall now consider, we take $H_1 = I$ and w_i a multiple of z_i, implying, from the idempotency of H_i, that

$$H_{i+1} = H_i - \frac{H_i A^T p_i p_i^T}{p_i^T A p_i} \, . \tag{8.24}$$

Definition 8.5

The subclass of the ABS class of conjugate direction algorithms where $H_1 = I$ and w_i is a multiple of z_i is called the restricted ABS class of conjugate direction algorithms.

If in the restricted ABS class of conjugate direction algorithms we take $z_i = e_i$, we obtain an algorithm that can be interpreted as the implicit LU algorithm applied to the matrix $P^T A$, with P upper triangular. We have the following.

Theorem 8.7

The algorithm in the restricted ABS class of conjugate direction algorithms where $z_i = e_i$ is well defined and is equivalent to the implicit LU algorithm in the sense of generating the same set p_i and x_i.

Proof

For well-definiteness it is enough to show that $p_i \neq 0$, which is true since the ith component of p_i is one, independently of the coefficient matrix. Since $V = P$ is upper triangular, the equivalence with the LU algorithm follows from Theorems 7.19 and 7.22. Q.E.D.

Note that from the properties of the implicit LU factorization algorithm the search vectors generated by the above algorithm are not only A-conjugate, but also $P^T A$-semiconjugate. The number of multiplications required by the algorithm is no more than $(5/6)n^3 + O(n^2)$, $n^3/2$ multiplications coming from the evaluation of Ap_i.

Two classical algorithms generating A-conjugate directions are the Hestenes–Stiefel (1952) method and the Lanczos (1952) method. These algorithms belong to the restricted ABS class of conjugate direction algorithms. The Hestenes–Stiefel method is given by the following procedure.

ALGORITHM 12:   The Hestenes–Stiefel Conjugate Gradient Method
(A12)   Let $x_1 \in R^n$ be arbitrary. Compute the residual $r_1 = Ax_1 - b$. Stop if $r_1 = 0$, otherwise set $p_1 = r_1$ and $i = 1$.
(B12)   Update x_i by

$$x_{i+1} = x_i - \frac{p_i^T r_i}{p_i^T A p_i} p_i \ . \tag{8.25}$$

(C12)   Compute the residual r_{i+1}. Stop if $r_{i+1} = 0$.
(D12)   Compute the search vector p_{i+1} by

$$p_{i+1} = r_{i+1} - \frac{p_i^T A r_{i+1}}{p_i^T A p_i} p_i \ . \tag{8.26}$$

(E12)   Increment the index i by one and go to (B12).

Among the properties of the method we recall the following (Hestenes and Stiefel 1952, Hestenes 1980).

— The residuals are mutually orthogonal, $r_i^T r_j = 0$, $i \neq j$.
— The residual r_{i+1} is orthogonal to $p_1, \ldots, p_i$.
— Range$(P^i) = \text{Span}(r_1, \ldots, r_i) = \text{Span}(r_1, \ldots, A^{i-1} r_1)$.
— The index i at which termination occurs is equal to the dimension of the Krylov space $K(r_1, A)$ and is therefore no greater than the number of distinct eigenvalues of A.
— The function $F(x)$ given by (8.19) is minimized by x_{i+1} along the linear variety $x_1 + \text{Span}(p_1, \ldots, p_i)$. Moreover the following inequality is true:

$$F(x_{i+1}) - F(x^+) \leq 2[F(x_i) - F(x^+)] \frac{\text{Cond}(A) - 1}{\text{Cond}(A) + 1} \ . \tag{8.27}$$

The Hestenes–Stiefel method is a member of subclass S2, as shown by the following theorem.

Theorem 8.8
Consider the algorithm in the restricted ABS class of conjugate direction algorithms where z_i is given by

$$z_i = r_i \ . \tag{8.28}$$

Then the algorithm is well defined for all indexes such that r_i is nonzero and the sequences x_i, p_i are identical with those generated by the Hestenes–Stiefel method with the same starting point.

The proof of Theorem 8.8 is omitted, since it is a special case of Theorem 8.29. A proof similar to that of Theorem 8.5 is given by Abaffy and Spedicato (1985).

The Lanczos method is based upon the following procedure:

ALGORITHM 13: The Lanczos Conjugate Gradient Method

(A13) Let $\bar{x}_1 \in R^n$ be arbitrary. Compute the residual $\bar{r}_1 = A\bar{x}_1 - b$. Stop if $\bar{r}_1 = 0$, otherwise set $\bar{p}_1 = r_1$, $p_0 = 0$ and $i = 1$.

(B13)   Update the estimate of the solution by

$$\bar{x}_{i+1} = \bar{x}_i - \frac{\bar{r}_i^T \bar{p}_i}{\bar{p}_i^T A \bar{p}_i} \, \bar{p}_i \; . \tag{8.29}$$

(C13)   Compute the residual $\bar{r}_{i+1}$. Stop if $\bar{r}_{i+1} = 0$.

(D13)   Compute the search vector $\bar{p}_{i+1}$ by

$$\bar{p}_{i+1} = A\bar{p}_i - \frac{\bar{p}_i^T A^2 \bar{p}_i}{\bar{p}_i^T A \bar{p}_i} \, \bar{p}_i - \frac{\bar{p}_{i-1}^T A^2 \bar{p}_i}{\bar{p}_i^T A \bar{p}_i} \, \bar{p}_{i-1} \tag{8.30}$$

(E13)   Update i by one and go to (B13).

The search vectors generated by the Lanczos method are multiples of those generated by the Hestenes–Stiefel method and thus the two methods have essentially the same properties. This fact is related to a general result on conjugate directions methods (see for instance Oren (1972)) which says that search vectors that are A-conjugate and have moreover the property $\text{Span}(p_1, \ldots, p_i) = \text{Span}(r_1, \ldots, r_i)$ are uniquely determined up to a scaling factor (see also Remark 7.9). Note that there are alternative implementations of the Lanczos method (see for instance Paige and Saunders (1975), Chandra (1978), Parlett (1980), O'Leary (1980), Simon (1982)), which require only symmetry (and not positive definiteness).

The parameter choices in the ABS class generating an algorithm equivalent to the Lanczos method are given in Theorem 8.9, where use is made of the following two lemmas.

Lemma 8.2

Consider the scaled ABS class with $H_1 = I$. Suppose that $m = n$, that A is nonsingular and that the scaling vectors satisfy

$$v_i = A^{-T} \sum_{k=1}^{i+1} \tau_{i,\,k} \tau_k \; , \tag{8.31}$$

where the $\tau_{i,\,k}$ are scalars. Then, if the residuals satisfy a relation of the form

$$r_i^T G r_j = 0 , \quad i < j , \tag{8.32}$$

where G is some matrix in $R^{n,n}$, it follows that

$$H_i^T G r_j = G r_j , \quad i < j . \tag{8.33}$$

Proof
For $i = 1$ the lemma is trivial. Proceeding by induction, we suppose the lemma true up to i and we prove it for $j > i + 1$. We have $H_{i+1}^T G r_j = H_i^T G r_j - H_i^T w_i v_i^T A H_i^T G r_j$. By the induction $v_i^T A H_i^T G r_j = v_i^T A G r_j$. Using (8.31), we have

$$v_i^T A G r_j = \sum_{k=1}^{i+1} \tau_{i,k} \tau_k^T G r_j \tag{8.34}$$

which is zero owing to (8.32), completing the induction. Q.E.D.

Remark 8.4
Equation (8.33) is satisfied by all algorithms in the restricted ABS class of conjugate directions, provided that the step size is nonzero. Indeed we have

$$r_{i+1} = r_i - \alpha_i A p_i . \tag{8.35}$$

Hence (8.31) holds with $\tau_{i,k} = 0$ for $k < i$, $\tau_{i,i} = 1/\alpha_i$, $\tau_{i,i+1} = -1/\alpha_i$. For the Hestenes–Stiefel and the Lanczos methods the residuals are orthogonal; hence (8.32) holds with $G = I$.

Lemma 8.3
Let A be symmetric positive definite and consider the restricted ABS class of conjugate direction algorithms. Then the following relation is true:

$$H_i A p_j = A p_j , \quad j \geq i . \tag{8.36}$$

Proof
For $i = 1$ the lemma is trivial. Proceeding by induction, we assume the lemma to be true up to i and we prove it for $i + 1$. Using (8.24) we have, for $j \geq i + 1$,

$$H_{i+1} A p_j = H_i A p_j - \frac{H_i A p_i p_i^T A p_j}{p_i^T A p_i} . \tag{8.37}$$

The first term on the right-hand side of (8.37) equals $A p_j$ by induction. In the second term we have $p_i^T A p_j = v_i^T A p_j = z_j^T H_j A v_i$, which is zero, since $A v_i$ lies in the null space of H_j for $j > i$. Q.E.D.
   We can now prove the following.

Theorem 8.9
Consider the algorithm in the restricted ABS class of conjugate directions where $z_1 = r_1$ and, for $i > 1$,

$$z_i = Ap_{i-1} \; . \tag{8.38}$$

Then the algorithm is well defined for all indexes such that r_i is nonzero and the sequences x_i, p_i are identical with the sequences $\bar{x}_i$, $\bar{p}_i$ generated by the Lanczos method with the same starting point $x_1 = \bar{x}_1$.

Proof
Since the formulas for the step size are identical for the two algorithms, it is enough to show that the sequences p_i, $\bar{p}_i$ are identical. By assumption $p_1 = \bar{p}_1$. From (8.24) we have

$$H_{i+1}{}^{\mathrm{T}} = H_i{}^{\mathrm{T}} - \frac{p_i p_i{}^{\mathrm{T}} A H_i{}^{\mathrm{T}}}{p_i{}^{\mathrm{T}} A p_i} \; . \tag{8.39}$$

Hence $p_2 = H_2{}^{\mathrm{T}} z_2 = Ap_1 - (p_1{}^{\mathrm{T}} A^2 p_1 / p_1{}^{\mathrm{T}} A p_1) p_1 = p_2$. Also we have

$$p_3 = \left(I - \frac{p_1 p_1{}^{\mathrm{T}} A}{p_1{}^{\mathrm{T}} A p_1} - \frac{p_2 p_2{}^{\mathrm{T}} A H_2{}^{\mathrm{T}}}{p_2{}^{\mathrm{T}} A p_2} \right) A p_2 \; . \tag{8.40}$$

Using Lemma 8.3 twice we get $p_2{}^{\mathrm{T}} A H_2{}^{\mathrm{T}} A p_2 = p_2{}^{\mathrm{T}} A H_2 A p_2 = p_2{}^{\mathrm{T}} A^2 p_2$; hence $p_3 = \bar{p}_3$. We proceed now by induction, assuming the theorem to be true up to i. We also recall that the residuals generated by the Lanczos method are orthogonal. Using (8.39) twice, p_{i+1} reads

$$p_{i+1} = \left(H_{i+1} - \frac{p_{i-1} p_{i-1}{}^{\mathrm{T}} A H_{i-1}{}^{\mathrm{T}}}{p_{i-1}{}^{\mathrm{T}} A p_{i-1}} - \frac{p_i p_i{}^{\mathrm{T}} A H_i{}^{\mathrm{T}}}{p_i{}^{\mathrm{T}} A p_i} \right) A p_i \; . \tag{8.41}$$

Using again Lemma 8.3, we get $p_{i-1}{}^{\mathrm{T}} A H_{i-1}{}^{\mathrm{T}} p_i = p_{i-1}{}^{\mathrm{T}} A^2 p_i$ and $p_i{}^{\mathrm{T}} A H_i{}^{\mathrm{T}} A p_i = p_i{}^{\mathrm{T}} A^2 p_i$; hence (8.41) gives

$$p_{i+1} = H_{i-1}{}^{\mathrm{T}} A p_i - \frac{p_i{}^{\mathrm{T}} A^2 p_i}{p_i{}^{\mathrm{T}} A p_i} p_i - \frac{p_{i-1}{}^{\mathrm{T}} A^2 p_i}{p_{i-1}{}^{\mathrm{T}} A p_{i-1}} p_{i-1} \; . \tag{8.42}$$

Thus $p_{i+1} = \bar{p}_{i+1}$ if and only if $H_{i-1}{}^{\mathrm{T}} A p_i = A p_i$. By Remark 8.4 and the orthogonality of the residuals in the Lanczos method, we can apply Lemma 8.2, with $G = I$. The identity $H_{i-1}{}^T Ap_i = Ap_i$ follows by writing (8.33) for $j = i + 1$ and $j = i$, subtracting the two relations and applying (8.35). Q.E.D.

For all algorithms in S2 the matrix $D = P^T AP$ is diagonal. The following theorem shows that for some algorithms in the restricted ABS class of conjugate directions algorithms, including the Lanczos and the Hestenes–Stiefel methods, the matrix $P^T A^2 P$ is tridiagonal.

Theorem 8.10

Let A be symmetric and positive definite and consider the restricted ABS class of conjugate direction algorithms. Assume that Lemma 8.2 applies with $G = I$. Then, if $m \geqslant 3$ and at least three iterations are made by the algorithm

$$P^T A^2 P = T \, , \tag{8.43}$$

where T is a symmetric and positive definite tridiagonal matrix. Moreover, if n iterations are made by the algorithm, then the following relation holds with D a diagonal matrix:

$$P^T P = DT^{-1}D \, . \tag{8.44}$$

Proof

Let $i \geqslant 3$ be the termination index. To prove (8.43) it is enough to show that $p_j{}^T A^2 p_i = 0$ for $j \leqslant i - 2$. From Lemma 8.2 we have that $H_{i-1}{}^T Ap_i = Ap_i$. From (8.24) we also get

$$H_{i-1}{}^T = I - \sum_{j=1}^{i-2} \frac{p_j p_j{}^T A H_j{}^T}{p_j{}^T A p_j} \, . \tag{8.45}$$

Multiplying (8.45) on the right by Ap_i we get

$$\sum_{j=1}^{i-2} \frac{p_j{}^T A H_j Ap_i}{p_j{}^T A p_j} p_j = 0 \, . \tag{8.46}$$

By Lemma 8.3 $H_j Ap_i = Ap_i$; hence $p_j{}^T A^2 p_i = 0$ from the linear independence of the search vectors. If $i = n$ then $P^T AP = D$ is diagonal, implying that $AP = P^{-T}D$, $P^T A = DP^{-1}$; hence (8.44) follows by substituting in (8.43). Q.E.D.

8.4 SUBCLASS S4; THE ORTHOGONALLY SCALED ABS ALGORITHM

In this section we assume that A is square nonsingular and we consider the subclass S3 of the ABS class where v_i is given by

$$v_i = Ap_i \, . \tag{8.47}$$

The well-definiteness condition (7.6) is satisfied, since it reads $v_i^T A p_i = p_i^T A^T A p_i > 0$. Choice (8.47) corresponds to taking $Y = A^T A$ and $\delta_i = 1$ in (7.70); hence we can restate the results in Theorem 7.17 as the following.

Theorem 8.11
Let A be square nonsingular. Then the scaled ABS class where $v_i = A p_i$ generates $A^T A$-conjugate search vectors and x_{i+1} minimizes, over the linear variety $x_1 + \text{Span}(p_1, \ldots, p_i)$, the quadratic function

$$F(x) = (x - x^+)^T A^T A (x - x^+) \tag{8.48}$$

or

$$F(x) = r(x)^T r(x) , \tag{8.49}$$

where $r(x) = Ax - b$ is the residual vector.

Before giving another variational characterization of subclass S3, we prove the following.

Theorem 8.12
For any algorithm in the ABS class the residuals satisfy

$$r_{i+1} = (I - R_i) r_i , \tag{8.50}$$

where R_i is the idempotent rank-one matrix given by

$$R_i = \frac{A p_i v_i^T}{v_i^T A p_i} , \tag{8.51}$$

and R_i satisfies the left orthogonality relations

$$R_j R_i = 0 , \quad j < i . \tag{8.52}$$

Proof
Equations (7.7) and (7.8) imply that $r_{i+1} = r_i - A p_i v_i^T r_i / v_i^T A p_i$, yielding (8.50) and (8.51). The idempotency of R_i is immediately verified. Now $R_j R_i = A p_j v_j^T [v_j^T A p_i / (v_j^T A p_j v_i^T A p_i)]$ and (8.52) follows since $v_j^T A p_i = p_i^T A^T v_j = z_i^T H_i A^T v_j$ is zero for $j < i$ from Theorem 7.2. Q.E.D.

Theorem 8.13
The Frobenius norm of R_i is globally minimized and is equal to one, if and only if the scaling vector v_i satisfies (8.47). Moreover, for such a choice, R_i is symmetric and

$$R_j R_i = 0 , \qquad j > i .$$ (8.53)

Proof
Since we have the identity

$$\|R_i\|_F^2 = \frac{\|v_i\|_2^2 \|Ap_i\|_2^2}{(v_i^T Ap_i)^2} ,$$ (8.54)

the minimality property of R_i follows from the Cauchy–Schwartz inequality. Symmetry of R_i follows from (8.47) and (8.51), while (8.53) follows from (8.52). Q.E.D.

It is easy to verify that the reduction in the each step in function $F(x)$ defined in (8.49) is given by

$$F(x_i) - F(x_{i+1}) = \frac{(r_i^T v_i)^2}{v_i^T v_i} .$$ (8.55)

An interesting question is which choice of z_i maximizes the decrease in the residual norm. We have the following.

Theorem 8.14
In subclass S3 the decrease in the residual norm is maximized, and $x_{i+1} = x^+$, if z_i satisfies the compatible equation

$$H_i^T z_i = A^{-1} r_i .$$ (8.56)

Proof
From the Cauchy–Schwartz inequality the right hand side in (8.55) is maximized for $v_i = Ap_i = r_i$, giving

$$F(x_i) - F(x_{i+1}) = r_i^T r_i .$$ (8.57)

Hence, from the definition of F, $F(x_{i+1}) = 0$ or $x_{i+1} = x^+$. To establish the compatibility of (8.56), just observe that the right hand side is orthogonal to $A^T v_j$ for $j < i$ and that the set of vectors of the form $H_i^T z_i$ is the orthogonal complement of Range($A^T V^{i-1}$). Finally note that condition (7.6) is satisfied, since $z_i^T H_i A^T v_i = r^{iT} r_i) > 0$.

We give now some other properties of the considered subclass.

Theorem 8.15
The scaling vectors defined by (8.47) are mutually orthogonal.

Proof
We have $v_i^T v_j = p_i^T A^T v_j = z_i^T H_i^T v_j = 0$ for $i > j$ from Theorem 7.2.

Definition 8.6
The subclass of the scaled ABS class where v_i is given by (8.47) is called the class of the orthogonally scaled ABS algorithms.

Theorem 8.16
If the scaling vectors are given by (8.47) and termination requires n iterations, then the inverse A^{-1} has the form

$$A^{-1} = \sum_{j=1}^{n} \frac{p_j v_j^T}{v_j^T v_j} .$$

(8.58)

Proof
Repeated application of (8.50) gives $r_{n+1} = (I - R_n)(I - R_{n-1})...(I-R_1)r_1$. From (8.52) we obtain

$$(I - R_n)(I - R_{n-1})...(I - R_1) = I - \sum_{j=1}^{n} R_j .$$

(8.59)

Since r_{n+1} is zero for arbitrary r_1, we have $I - \sum_{j=1}^{n} R_j = 0$, implying from (8.51) and (8.47) that

$$I = A \sum_{j=1}^{n} \frac{p_j v_j^T}{v_j^T v_j}$$

(8.60)

and the theorem follows. Q.E.D.

Remark 8.6
From (8.58) we can write the solution in the form

$$x^+ = \sum_{j=1}^{n} \frac{v_j^T b}{v_j^T v_j} p_j \; .$$

(8.61)

If $x_1 = 0$, the solution also has the form $x^+ = \sum_{j=1}^{n} - \alpha_j p_j$. Thus we obtain for the step size

$$\alpha_j = \frac{-v_j^T b}{v_j^T v_j} \; .$$

(8.62)

Hence the algorithms in subclass S3 can be implemented without explicit evaluation of the residual.

Theorem 8.17
Consider the orthogonally scaled ABS class with $H_1 = I$ and w_i a multiple of z_i. Then the update of H_i can be written in the form

$$H_{i+1} = H_i - \frac{H_i A^T v_i p_i^T}{v_i^T v_i}$$

(8.63)

or

$$H_{i+1} = I - \sum_{j=1}^{i} \frac{A^T v_j p_j^T}{v_j^T v_j} \; .$$

(8.64)

Moreover the search and the scaling vectors have the form

$$p_{i+1} = z_{i+1} - \sum_{j=1}^{i} \frac{v_j^T A z_{i+1}}{v_j^T v_j} p_j$$

(8.65)

$$v_{i+1} = A z_{i+1} - \sum_{j=1}^{i} \frac{v_j^T A z_{i+1}}{v_j^T v_j} v_j \; .$$

(8.66)

Proof
Condition (7.10) is satisfied when w_i has the form

$$w_i = \frac{u_i}{u_i^T H_i A^T v_i}$$

(8.67)

for some u_i not orthogonal to $H_i A^T v_i$. Thus, when w_i and z_i are proportional, we can take

$$w_i = \frac{z_i}{z_i^T H_i A^T v_i} \cdot \tag{8.68}$$

Hence (8.63) follows from (7.9) by the definition of p_i and v_i. Equation (8.64) is easily obtained by induction, and (8.65) and (8.66) follow from the definition of p_i and v_i.

Definition 8.7
The subclass of the orthogonally scaled ABS class where $H_1 = I$ and w_i is a multiple of z_i is called the restricted orthogonally scaled ABS class.

In the restricted orthogonally scaled ABS class there exists an algorithm which generates an implicit factorization of A into the product of an orthogonal by an upper triangular matrix. This algorithm can therefore be considered as the iterative version of the classical QR algorithm.

Theorem 8.18
Consider the algorithm of the restricted orthogonally scaled ABS class where z_i is proportional to e_i. Then this algorithm is well defined and moreover, if n iterations are required for termination, it implicitly factorizes A into the product $A = QR$, Q orthogonal and R upper triangular.

Proof
For all algorithms in subclass S3 which need n iterations for termination, we have from (7.14)

$$A = V^{-T} D' P^{-1} . \tag{8.69}$$

where $V = V^n$, $P = P^n$ and $D' = L^n$ is diagonal. Since the columns of V are orthogonal, we can write $V = Q\overline{D}$, where Q is an orthogonal matrix and $\overline{D}$ is diagonal; hence (8.69) reads

$$A = QDP^{-1} \tag{8.70}$$

with $D = \overline{D}^{-1} D'$. Now the given choice of z_i implies that we are considering the implicit LU algorithm applied to the problem with coefficient matrix $V^T A$. Thus, if the algorithm is well defined, P^{-1} is upper triangular and the theorem follows, with $R = DP^{-1}$. The implicit LU algorithm is well defined without pivoting if and only if it is applied to a strongly nonsingular matrix. Now $V^T A = P^T A^T A$ consists of the positive definite matrix $A^T A$ premultiplied by the nonsingular lower triangular matrix P^T. Since a strongly nonsingular matrix remains strongly nonsingular if it is premultiplied by a nonsingular lower triangular matrix, it follows easily by induction that V^TA is strongly nonsingular. Thus the algorithm is well defined and the theorem is proved. Q.E.D.

Definition 8.8
The algorithm of the restricted orthogonally scaled ABS class where z_i is proportional to e_i is called the implicit QR algorithm.

Remark 8.7
If the implicit QR algorithm is implemented using the general formulas for the scaled ABS algorithm, it is easily seen that the required number of multiplications is $(11/6)n^3 + O(n^2)$. If (8.65) and (8.66) are used, there is a reduction in the overhead, the leading term being now $(4/3)n^3$ (or $m^2n + m^3/3$ for $m < n$). For other implementations of the implicit QR algorithm based upon the alternative formulations of the ABS algorithm, see Bodon (1989).

Remark 8.8
If $m < n$, A^TA is positive semidefinite and V^TA may not be strongly nonsingular. If $m > n$ and A is full rank, then A^TA is positive definite and the first n principal submatrices of V^TA are nonsingular. Thus the implicit QR algorithm is well defined on overdetermined full rank systems (for the first n steps, since $H_{n+1} = 0$). In the next chapter we show that in such a case x_{n+1} is the least squares solution.

8.5   SUBCLASS S4: THE CLASS OF THE OPTIMALLY STABLE ABS ALGORITHMS

In this section we assume that A is square nonsingular and we consider the subclass S4 of the ABS class where V_i is defined by

$$V_i = A^{-T}p_i \ . \tag{8.71}$$

The well-definiteness condition (7.6) is satisfied, since it reads $p_i^Tp_i > 0$. For the moment we do not discuss the computational problems due to the presence of the inverse.

Choice (8.71) corresponds to taking $Y = I$ and $\delta_i = 1$ in (7.70); hence we can restate the results in Theorem 7.17 as the following.

Theorem 8.19
Let A be square nonsingular. Then the scaled ABS class where $v_i = A^{-T}p_i$ generates orthogonal search vectors and x_{i+1} minimizes over the linear variety $x_1 + \text{Span}(p_1, \ldots, p_i)$ the convex quadratic function

$$F(x) = (x - x^+)^T(x - x^+) \tag{8.72}$$

i.e. the Euclidean distance from the solution.

Before giving another variational characterization of subclass S4, we give the following.

Theorem 8.20
Define the error $s_i \in R^n$ in the approximation to the solution by

$$s_i = x_i - x^+ . \tag{8.73}$$

Then for any algorithm in the scaled ABS class the errors satisfy

$$s_{i+1} = (I - S_i)s_i , \tag{8.74}$$

where S_i is the idempotent rank-one matrix given by

$$S_i = \frac{p_i v_i^T A}{v_i^T A p_i} , \tag{8.75}$$

which satisfies the left orthogonality relation

$$S_j S_i = 0 , \quad j < i . \tag{8.76}$$

Proof
By subtracting x^+ from both sides in (7.7) we get $s_{i+1} = s_i - p_i(v_i^T r_i / v_i^T A p_i)$. From the identity $r_i = A s_i$, (8.74) and (8.75) follow easily. Idempotency is immediately verified and (8.76) follows from Theorem 8.2. Q.E.D.

The proof of the following theorem is omitted, since it is similar to the proof of Theorem 8.12.

Theorem 8.21
The Frobenius norm of S_i is globally minimized and is equal to one, if and only if the scaling vector v_i satisfies (8.70). Moreover, for such a choice, S_i is symmetric and

$$S_j S_i = 0 , \quad j > i . \tag{8.77}$$

Choice (8.71) implies that $p_i = A^T v_i$, hence substituting in the implicit factorization relation (7.15) and taking into account that the left-hand side is symmetric, we have

$$p^T p = D' . \tag{8.78}$$

implying the orthogonality of the search vectors, or also

$$V^T A A^T V = D \; , \tag{8.79}$$

with D', D'' diagonal. Equation (8.79) is discussed in Chapter 12. It was found by Broyden (1985) as the condition characterizing the optimally stable algorithms in ALGORITHM 10, and therefore in the scaled ABS class, which is equivalent to ALGORITHM 10. We can therefore state the following.

Theorem 8.22
The algorithms of subclass S4 satisfy the Broyden optimality condition, and generate orthogonal search vectors.

Definition 8.8
The subclass S4 is called the class of the optimally stable ABS algorithms.

By proceeding as in Theorem 8.16 we can prove the following.

Theorem 8.23
If the scaling vectors v_i are given by (8.71), then the inverse has the form

$$A^{-1} = \sum_{j=1}^{n} \frac{p_j v_j^T}{p_j^T p_j} \; . \tag{8.80}$$

We now discuss the problem of the presence of the inverse in the definition of the scaling vector, showing that we can avoid its actual evaluation.

Let us first consider the update of H_i. As observed before, condition (7.10) can be satisfied by taking w_i as in (8.67). Then the update of H_i becomes independent of v_i and reads

$$H_{i+1} = H_i - \frac{H_i p_i u_i^T H_i}{u_i^T H_i p_i} \; . \tag{8.81}$$

A further simplification is gained if $H_1 = I$ and w_i is a multiple of z_i, a feasible choice since $z_i^T H_i A^T v_i = p_i^T p_i$ is nonzero. Then (8.81) gives

$$H_{i+1} = H_i - \frac{H_i p_i p_i^T}{p_i^T p_i} \tag{8.82}$$

or also, from idempotency of H_i,

$$H_{i+1} = H_i - \frac{H_i p_i p_i^T H_i}{p_i^T p_i} \; . \tag{8.83}$$

From (8.83) we observe that, if H_1 is symmetric, the sequence H_i consists of symmetric matrices. Moreover, if $H_1 = I$, we have the identity $H_i p_i = H_i H_i^T z_i = H_i^2 z_i = H_i z_i = H_i^T z_i = p_i$, implying that we can write (8.83) in the form (8.6) or also

$$H_{i+1} = H_i - \frac{(H_i p_i)(H_i p_i)^T}{(H_i p_i)^T (H_i p_i)} \; . \tag{8.84}$$

Also note that (8.83) implies that

$$H_{i+1} = H_1 - \sum_{j=1}^{i} \frac{p_j p_j^T}{p_j^T p_j} \tag{8.85}$$

and

$$p_{i+1} = H_1^T z_{i+1} - \sum_{j=1}^{i} \frac{p_j^T z_{i+1}}{p_j^T p_j} p_j \; . \tag{8.86}$$

While the update formula for H_i derived above is independent of v_i, the formula for the step size is apparently not, since it reads

$$\alpha_i = \frac{r_i^T A^{-T} p_i}{p_i^T p_i} \; . \tag{8.87}$$

In Theorem (8.24) we show that, if $H_1 = I$ and w_i is a multiple of z_i, then the inverse can be removed in the formula for the step size. First we give the following.

Definition 8.9
The subclass of the optimally stable ABS algorithms where $H_1 = I$ and w_i is a multiple of z_i is called the restricted class of optimally stable ABS algorithms.

Theorem 8.24
Consider the restricted class of optimally scaled ABS algorithms. Define z_i by

$$z_i = A^T u_i \; , \tag{8.88}$$

with u_i such that $p_i = H_i z_i = H_i A^T u_i \neq 0$. Then the step size is given by

$$\alpha_i = \frac{r_i^T u_i}{p_i^T p_i} \; . \tag{8.89}$$

Proof
For the above choices, equations (8.84) and (8.85) hold with $H_1 = I$, so that the search vectors are $p_1 = A^T u_1$ and for $i > 1$

$$p_{i+1} = A^T u_{i+1} - \sum_{j=1}^{i} \frac{p_j^T A^T u_{i+1}}{p_j^T p_j} p_j \ .$$

(8.90)

Recall also the following relation between the residuals:

$$r_{i+1} = r_i - \alpha_i A p_i \ .$$

(8.91)

Equation (8.89) follows if we show that $r_i^T u_i = r_i^T v_i = r_i^T A^{-T} p_i$. We proceed by induction. For $i = 1$ the result is immediate, since $p_1 = A^T u_1$. Assuming now (8.89) to be true up to i, using (8.91) and the orthogonality of the search vectors (see Theorem 8.22), we obtain $r_{i+1}^T v_{i+1} = (r_i - \alpha_i A p_i)^T A^{-T} p_{i+1} = r_i^T A^{-T} p_{i+1}$. Using (8.90), we have now $r_i^T A^{-T} p_{i+1} = r_i^T u_{i+1} - \sum_{j=1}^{i} (p_j^T A^T u_{i+1}/p_j^T p_j) r_i^T A^{-T} p_j$. Consider the term $r_i^T A^{-T} p_j$ in the above summation. For $j = i$ we have $r_i^T A^{-T} p_i = r_i^T v_i = \alpha_i p_i^T p_i$ by induction. For $j < i$ we have $r_i^T A^{-T} p_j = r_i^T v_j = 0$ from Theorem 7.5. Thus $r_i^T A^{-T} p_{i+1} = (r_i - \alpha_i A p_i)^T u_{i+1} = r_{i+1}^T u_{i+1}$, completing the induction. Q.E.D.

Remark 8.10
Since A is nonsingular, there is no loss of generality in definition (8.87).

The following theorem gives an additional characterization of the Huang algorithm.

Theorem 8.25
The Huang algorithm and the algorithm of the restricted class of optimally stable ABS algorithms where $u_i = e_i$ generate the same iterates.

Proof
The proof is immediate by induction.

Theorem 8.25 is a special case of the following theorem.

Theorem 8.26
The set of sequences p_i, x_i and $\bar{p}_i$, $\bar{x}_i$ generated respectively by the algorithms in subclass S1 and S4 are identical.

Proof
Without loss of generality we can assume for both classes that $H_1 = I$. The search vectors in subclass S1 have the form $p_i = H_i A^T v_i$, with H_i updated by $H_{i+1} = H_i - p_i p_i^T / p_i^T p_i$. The search vectors in subclass S4 have the form $\bar{p}_i = \bar{H}_i A^T u_i$, with $\bar{H}_i$ updated by $\bar{H}_{i+1} = \bar{H}_i - \bar{H}_i \bar{p}_i q_i^T \bar{H}_i / q_i^T \bar{H} \bar{p}_i$. Without loss of generality we can take $q_i = \bar{p}_i$, hence $\bar{H}_i$ is symmetric and $\bar{H}_i \bar{p}_i = \bar{p}_i$. By induction it follows immediately that $p_i = \bar{p}_i$ and $H_i = \bar{H}_i$ if the sequences u_i, v_i are identical. To establish the identity of the sequences x_i, $\bar{x}_i$, it is now enough to show that, when $x_1 = \bar{x}_1$, the step sizes are

identical. In subclass S1 the step size has the form $\alpha_i = r_i^T v_i / p_i^T A^T v_i = r_i^T v_i / p_i^T p_i$, since $p_i^T A^T v_i = v_i^T A H_i A^T v_i = v_i^T A H_i^2 A^T v_i = p_i^T p_i$. If $v_i = u_i$, we have $p_i = \bar{p}_i$, hence the identity $\alpha_i = \bar{\alpha}_i$ follows immediately by induction using (8.89). Q.E.D.

Theorem 8.27
The vectors u_i, p_i in the restricted class of optimally stable ABS algorithms are A-semiconjugate, i.e.

$$u_j^T A p_i = 0 , \qquad 1 \leqslant j < i \leqslant n . \tag{8.92}$$

Proof
From (8.89) we get

$$A^T u_j = p_j + \sum_{k=1}^{j-1} \delta_k p_k , \tag{8.93}$$

with $\delta_k = p_k^T A^T u_j / p_k^T p_k$. From (8.93) we get $u_j^T A p_i = p_i^T A^T u_j = p_i^T \left(p_j + \sum_{k=1}^{j-1} \delta_k p_k \right) = 0$, by the orthogonality of the search vectors. Q.E.D.

We consider now the algorithm in the restricted optimally stable ABS class where $u_i = r_i$. Using (8.91), (8.90) reads

$$p_{i+1} = A^T r_{i+1} - \sum_{j=1}^{i} \frac{(r_{j+1} - r_j)^T r_{i+1}}{r_j^T r_j} p_j . \tag{8.94}$$

Since the (nonzero) step size α_j has the form

$$\alpha_j = \frac{r_j^T r_j}{p_j^T p_j} , \tag{8.95}$$

we can write (8.94) also as

$$p_{i+1} = A^T r_{i+1} - \sum_{j=1}^{i} \frac{(r_{j+1} - r_j)^T r_{i+1}}{r_j^T r_j} p_j \tag{8.96}$$

The following theorem establishes two remarkable properties of the algorithm given by (8.96), namely that the residuals are orthogonal and that only the last term in the summation can be nonzero. Hence the algorithm is of the conjugate gradient type.

Theorem 8.28

Define the search vectors by (8.96). Then the residuals are orthogonal and (8.96) can be written in the form

$$p_{i+1} = A^T r_{i+1} + \frac{r_{i+1}^T r_{i+1}}{r_i^T r_i} p_i \ . \tag{8.97}$$

Proof

We proceed by induction. From (8.90), (8.94) and the definition $p_1 = A^T r_1$ we have $r_2^T r_1 = [r_1 - (r_1^T r_1/r_1^T AA^T r_1)AA^T r_1]^T r_1 = 0$, so that the orthogonality is true for the first two vectors. Assuming now that the first i residuals are orthogonal, we must prove that $r_{i+1}^T r_j = 0$ for $j \le i$. For $j = i$, from (8.90) and (8.94) we have $r_{i+1}^T r_i = (r_i - r_i^T r_i/p_i^T p_i Ap_i)^T r_i = r_i^T r_i (1 - r_i^T Ap_i/p_i^T p_i)$. Using the induction, we obtain from (8.96)

$$p_i = A^T r_i + \frac{r_i^T r_i}{r_{i-1}^T r_{i-1}} p_{i-1} \tag{8.98}$$

or

$$A^T r_i = p_i - \frac{r_i^T r_i}{r_{i-1}^T r_{i-1}} p_{i-1} \ . \tag{8.99}$$

(8.98) and the orthogonality of p_i, p_{i-1} give $r_i^T Ap_i = p_i^T p_i$ so that, by substituting in the previous expression, we get $r_{i+1}^T r_i = 0$. For $j < i$, we have, using the induction, $r_{i+1}^T r_j = - (r_i^T r_i/p_i^T p_i)p_i^T A^T r_j$, which is zero from Theorem 8.27. Finally, (8.97) follows immediately from (8.96).

Remark 8.11

The conjugate gradient-type method defined by (8.98) goes back to Craig (1955) and is sometimes known as the method of minimum AA^T-iterations (Voyevodin 1983). Since the method is well defined under the only assumption of nonsingularity of A, it is of interest for the solution of large nonsymmetric methods (see for instance Elman (1982) and Radicati di Brozolo and Robert (1987)).

Remark 8.12

From (8.97) we derive easily

$$p_{i+1} = A^T \left[r_{i+1} - r_{i+1}^T r_{i+1} \left(\frac{r_i}{r_i^T r_i} + \ldots + \frac{r_1}{r_1^T r_1} \right) \right] \ . \tag{8.100}$$

Since by Theorem 8.28 the vectors r_i are orthogonal, it follows that the necessary and sufficient condition for p_{i+1} to be nonzero and linearly independent from $p_1, \ldots, p_i$ is that r_{i+1} is nonzero. Thus the algorithm can terminate for $i < n$ only at points which solve the system. If A is symmetric, then (8.100) implies that

$$p_{i+1} = \sum_{j=1}^{i+1} \mu_j A^j r_1 \tag{8.101}$$

for some scalars μ_j. It follows that the algorithm must terminate in a number of steps no greater than the dimension of the Krylov space $K(Ar_1, A)$, which is no greater than the number of the distinct eigenvalues of A.

8.6 SUBCLASS S5: THE VOYEVODIN CLASS OF CONJUGATE GRADIENT-TYPE ALGORITHMS

A class of conjugate gradient-type algorithms for nonsingular linear systems is presented by Voyevodin (1983). The class is available in two alternative forms, is formulated in terms of general inner products and contains as parameters two matrices B and C. The matrices B and C are nonsingular and satisfy

$$CABC^{-1} = \alpha I + \beta B^T A^T , \tag{8.102}$$

where α and β are some scalars. The first formulation of the class is given by the following procedure.

ALGORITHM 14: The Voyevodin First Class of Conjugate Gradients
(A14) Let $x_1 \in R^n$ be arbitrary. Compute $r_1 = Ax_1 - b$. If $r_1 = 0$, stop; otherwise set $s_1 = r_1, i = 1$.
(B14) Update the estimate of the solution by

$$x_{i+1} = x_i - \alpha_i B s_i \tag{8.103}$$

where

$$\alpha_i = \frac{(r_i, Cr_i)}{(r_i, CABs_i)} . \tag{8.104}$$

(C14) Compute the residual r_{i+1}. If $r_{i+1} = 0$, stop.
(D14) Compute the search vector s_{i+1} by

$$s_{i+1} = r_{i+1} + \beta_i ABs_i , \tag{8.105}$$

where

$$\beta_i = \frac{-(s_i, \, CABr_{i+1})}{(s_i, \, CABs_i)} \; . \tag{8.106}$$

(E14) Increment the index i by one and go to (B14).

The second class of Voyevodin is obtained from the first class by the substitution $u_i = Bs_i$. Note that, in ALGORITHM 14, alternative formulas can be used for α_i and β_i, e.g.

$$\alpha_i = \frac{(s_i, \, Cr_i)}{(s_i, \, CABs_i)} \; . \tag{8.107}$$

Sufficient conditions for ALGORITHM 14 to be well defined (no zero divisions) and not to stop short of the solution ($\alpha_i = 0$ when $r_i \neq 0$) are that C and CAB be symmetric positive definite. In such a case the algorithm has the following properties.

— It terminates after $p \leq n$ iterations, p being the dimension of the Krylov $K(r_1, \, AB)$, which is no greater than the number of distinct eigenvalues of AB. Note that B can be used to reduce the number of distinct eigenvalues.
— The search vectors are CAB-conjugate and the residuals are C-conjugate.
— The residual r_{i+1} is C-orthogonal to $s_1, \, \ldots, \, s_i$, i.e.

$$(s_k, \, Cr_{i+1}) = 0 \qquad k \leq i \; . \tag{8.108}$$

— Two step recursions hold for the search vectors and the residuals, namely

$$s_{i+1} = -\alpha_i ABs_i + (1 + \beta_i)s_i - \beta_{i-1}s_{i-1} \; , \tag{8.109}$$

$$r_{i+1} = -\alpha_i ABr_i + \left(1 + \frac{\beta_i \alpha_i}{\alpha_{i-1}}\right) r_i - \frac{\beta_i \alpha_i}{\alpha_{i-1}r_{i-1}} \; . \tag{8.110}$$

— Define the generalized error function $F(s)$, $s = x - x^+$, by

$$F(s) = (s, \, Rs) \; , \tag{8.111}$$

where R is the symmetric positive definite matrix given by

$$R = B^{-T}(CAB)B^{-1} \tag{8.112}$$

Then x_{i+1} minimizes $F(s)$ over the linear variety $x_1 + \text{Span}(s_1, \, \ldots, \, s_i)$.

Defining now the inner product $(a, \, b)$ by the scalar product $a^T b$, we consider some admissible choices in the Voyevodin first class.

— $B = C = I$: admissible if A is symmetric positive definite; this algorithm coincides with the Hestenes-Stiefel method (see section 8.2).

— $B = A^T$, $C = I$: admissible for A nonsingular; this algorithm coincides with the Craig method (see section 8.4).

— $B = A^T$, $C = AA^T$: admissible for A nonsingular; this algorithm is the so-called minimum $A^T A$-iterations algorithm, which can be obtained by applying the Hestenes–Stiefel method on the normal system $A^T Ax = A^T b$.

— Let A be nonsingular and consider the Hermitian decomposition of A into its symmetric and antisymmetric part, i.e.

$$A = M + N \ , \tag{8.113}$$

where $M = M^T$ and $N = -N^T$. Then, if M is nonsingular, condition (8.102) is satisfied, with $\alpha = 2$ and $\beta = -1$, by $B = C = M^{-1}$. The algorithm, called the method of complete Hermitian decomposition, is well defined if M is positive definite.

— Let A be symmetric positive definite and write it in the form (8.113) but with M and N symmetric. Then, if M is nonsingular, condition (8.102) is satisfied, with $\alpha = 0$ and $\beta = 1$, by $B = C = M^{-1}$. The algorithm, called the method of incomplete Hermitian decomposition, is well defined if M is positive definite.

The parameter choices in the scaled ABS class that generate an algorithm equivalent to an algorithm in the Voyevodin class are given by the following theorem.

Theorem 8.29
Let A, B, C be nonsingular matrices which satisfy (8.102) and which are admissible choices in the Voyevodin first class. Consider the algorithm of the scaled ABS class where $H_1 = I$, z_i and v_i are given by

$$z_i = Br_i \ , \tag{8.114}$$

$$v_i = C^T B^{-1} p_i \tag{8.115}$$

and w_i is proportional to z_i. Then, if $x_1 = \bar{x}_1$, the sequences x_i, α_i, p_i generated by the algorithm are identical with the sequences $\bar{x}_i$, $\bar{\alpha}_i$, $u_i = Bs_i$ generated by the corresponding algorithm in the Voyevodin first class.

Proof
We proceed by induction. For $i = 1$, we have $p_1 = Br_1 = Bs_1 = u_1$ and $\alpha_1 = r_1^T C^T B^{-1} p_1 / p_1^T A^T C^T B^{-1} p_1 = (Cr_1)^T s_1 / (CAp_1)^T s_1 = (s_1, \ Cr_1)/(s_1, \ CABr_1) = (s_1, \ Cr_1)/(s_1, \ CABs_1) = \bar{\alpha}_1$ by (8.107), implying that $x_2 = \bar{x}_2$. Let us suppose now that $p_j = Bs_j$ and $\alpha_j = \bar{\alpha}_j$ for j up to $i-1$, implying that $x_i = \bar{x}_i$. To prove that $p_i = Bs_i$ and $\alpha_i = \bar{\alpha}_i$, we have to use a property of the vectors $H_j^T Br_i$ for $j = 1, \ldots, i-1$. Using Theorem 7.7, the definition of the w_i and the induction, we have

$$H_j^T = I - R_{j-1}(S_{j-1}^T CAR_{j-1})^{-1} S_{j-1}^T CA \ , \tag{8.116}$$

where $R_{j-1} = (Br_1, \ldots, Br_{j-1})$ and $S_{j-1} = (s_1, \ldots, s_{j-1})$. From (8.116) we get

$$H_i^T Br_i = Br_i - R_{j-1}(S_{j-1}^T CAR_{j-1})^{-1} S_{j-1}^T CABr_i . \tag{8.117}$$

From (8.117), r_i is a linear combination of s_i and s_{i-1}; hence the second term in (8.117) is zero from the CAB-conjugacy of the s_j. Thus we obtain, for $j = 1, \ldots, i-1$,

$$H_j^T Br_i = Br_i . \tag{8.118}$$

We can now show that $p_i = Bs_i$. Using the given parameter choices and relation $p_{i-1} = Bs_{i-1}$ which follows from the induction, the update of H_{i-1} reads

$$H_i^T = H_{i-1}^T - \frac{p_{i-1}s_{i-1}^T CAH_{i-1}^T}{s_{i-1}^T CABs_{i-1}} . \tag{8.119}$$

Since $H_{i-1}^T Br_i = Br_i$ from (8.118), we obtain for $p_i = H_i^T Br_i$

$$p_i = Br_i - \frac{s_{i-1}^T CABr_i}{s_{i-1}^T CABs_{i-1}} p_{i-1} . \tag{8.120}$$

Again using the induction, (8.120) can also be written as

$$p_i = B(r_i + \beta_{i-1}s_{i-1}) \tag{8.121}$$

where β_{i-1} is given according to (8.106). Hence, by (8.105), $p_i = Bs_i$ follows. Finally, we have $\alpha_i = r_i^T C^T B^{-1} p_i / p_i^T A^T C^T B^{-1} p_i = r_i^T C^T s_i / p_i^T A^T C^T s_i = r_i^T C^T s_i / s_i^T B^T A^T C^T s_i = \bar{\alpha}_i$, completing the induction. Q.E.D.

We can now express some of the previously considered special algorithms in the Voyevodin first class in terms of the parameters of the ABS class.

The Hestenes–Stiefel method ($B = C = I$) corresponds to the choices $v_i = p_i$, $z_i = r_i$, w_i proportional to r_i, as already seen in section 8.3.

The minimum AA^T-iterations algorithm ($B = A^T$, $C = I$) corresponds to the choices $v_i = A^{-T}p_i$, $z_i = A^T r_i$, w_i proportional to z_i. It coincides with the conjugate gradient-type algorithm of Craig (see section 8.5).

The minimum $A^T A$-iterations algorithm ($B = A^T$, $C = AA^T$) corresponds to the choices $v_i = Ap_i$, $z_i = A^T r_i$ and w_i proportional to z_i. It is therefore a member of subclass S3.

The method of complete Hermitian decomposition corresponds to the choices $v_i = p_i$, $z_i = Mr_i$ and w_i proportional to z_i. It is therefore a member of subclass S2, which does not require for well-definiteness that A be symmetric positive definite.

8.7 SUBCLASS S6: THE HEGEDÜS—BODOCS CLASS OF BIORTHOGONALIZATION ALGORITHMS

Hegedüs and Bodocs (1982) have introduced general recursions, here indicated as the HB recursions, to generate, given a matrix $A \in R^{m,n}$ with rank$(A) = q$, biortho-gonal vectors $v_j \in R^m$, $u_j \in R^n$, i.e. vectors satisfying for some $k \geq 1$

$$v_j^T A u_j \neq 0 , \qquad j = 1, \ldots, k , \tag{8.122}$$

$$v_j^T A u_i = 0 , \qquad i \neq j , \quad 1 \leq i , \quad j \leq k . \tag{8.123}$$

Hegedüs and Bodocs show that k can attain but not exceed q.

If A is full rank and $m \leq n$, then, letting $V = (v_1, \ldots, v_m)$, $U = (u_1, \ldots, u_m)$, the pair (V, U) is an A-conjugate pair in the sense of Definition 7.3. Since an A-conjugate pair is also A-semiconjugate, Theorem 7.13 implies that all the pairs (V, U) generated by the HB recursions can be generated by the scaled ABS algorithm. In this section we present the HB recursions and we give their formulation in terms of parameter choices in the scaled ABS class. Some particular cases in the HB recursions, where the vectors v_j, u_j are generated by conjugate gradient-type recursions, are also considered.

The HB recursions are defined by the following procedure. Suppose that vectors v_j, u_j have been obtained for $j = 1, \ldots, i - 1$ satisfying (8.122) and (8.123). Then two additional vectors v_i, u_i are generated by the formulas

$$v_i = S_i^T s_i , \tag{8.124}$$

$$u_i = Q_i q_i , \tag{8.125}$$

where S_i, Q_i are projectors given by $S_1 = Q_1 = I$ and, for $i \geq 2$,

$$S_i = I - \sum_{j=1}^{i-1} \frac{A u_j v_j^T}{v_j^T A u_j} , \tag{8.126}$$

$$Q_i = I - \sum_{j=1}^{i-1} \frac{u_j v_j^T A}{v_j^T A u_j} . \tag{8.127}$$

The vectors $s_i \in R^m$, $q_i \in R^n$ are called the basis vectors and are arbitrary save for the condition

$$s_i^T S_i A Q_i q_i \neq 0 . \tag{8.128}$$

It is always possible to satisfy (8.128) by some s_i, q_i, unless $S_i A Q_i = 0$. In such a case, (8.126) and (8.127) imply that

$$A = \sum_{j=1}^{i-1} \frac{Au_j v_j^{\mathrm{T}} A}{v_j^{\mathrm{T}} A u_j} \ , \tag{8.129}$$

or also

$$A = AU(V^{\mathrm{T}}AU)^{-1}V^{\mathrm{T}}A \ , \tag{8.130}$$

where $V^{\mathrm{T}}AU$ is diagonal. Hence by the rank factorization theorem $i-1$ must equal the rank of A and $U(V^{\mathrm{T}}AU)^{-1}V^{\mathrm{T}}$ is a generalized inverse of A, a result due to Egervary and already given in Lemma 6.2.

The following theorem shows that, if the vectors v_i of the HB recursions are identified with the scaling vectors of the ABS class, then for a certain choice of H_1, z_i and w_i the vectors u_i become identical with the vectors p_i. The proof is available in the paper by Abaffy and Spedicato (1985) and is omitted for brevity.

Theorem 8.30
Consider the HB recursions with basis vectors s_i, q_i satisfying condition (8.128). Consider the following parameter choices in the scaled ABS class: $H_1 = I$, v_i and z_i given by

$$v_i = S_i^{\mathrm{T}} s_i \ , \tag{8.131}$$

$$z_i = Q_i q_i \tag{8.132}$$

and w_i a multiple of z_i. Then these parameter choices are well defined and moreover the following identity is true:

$$p_i = Q_i q_i \ . \tag{8.133}$$

Remark 8.13
Equations (8.132) and (8.133) imply that

$$H_i^{\mathrm{T}} z_i = z_i \ . \tag{8.134}$$

An important feature of the HB recursions is that, for certain choices of q_i and s_i, only one term is left in (8.126) and (8.127), thereby generating conjugate gradient-type recursions for u_i and v_i. Here we present two of these choices, which are discussed by Hegedüs and Bodocs (1982).

— HB recursions of the Hestenes–Stiefel type: Let C and K be arbitrary positive definite matrices, and τ_i and μ_i be nonzero scalars. Generate the basis vectors by the following formulas, with s_1 and q_1 arbitrary:

$$s_{i+1} = s_i - \frac{\mu_i s_i^{\mathrm{T}} C s_i}{v_i^{\mathrm{T}} A u_i} A u_i \; , \tag{8.135}$$

$$q_{i+1} = q_i - \frac{\tau_i q_i^{\mathrm{T}} K q_i}{v_i^{\mathrm{T}} A u_i} A^{\mathrm{T}} v_i \; . \tag{8.136}$$

Then (8.124) and (8.125) take the form

$$v_{i+1} = \mu_{i+1} C s_{i+1} + \frac{\mu_{i+1} s_{i+1}^{\mathrm{T}} C s_{i+1}}{\mu_i s_i^{\mathrm{T}} C s_i} v_i \tag{8.137}$$

$$u_{i+1} = \tau_{i+1} K q_{i+1} + \frac{\tau_{i+1} q_{i+1}^{\mathrm{T}} K q_{i+1}}{\tau_i q_i^{\mathrm{T}} K q_i} u_i \; . \tag{8.138}$$

When A is symmetric, $s_1 = q_1$ and $\tau_i = \mu_i$, the above formulas yield $s_i = q_i$ and $v_i = u_i$. If moreover $K = C = I$, $\tau_i = \mu_i = 1$ and $s_1 = r_1 = A x_1 - b$, then the vectors s_i, v_i coincide with the vectors r_i, p_i of the Hestenes–Stiefel recursions. Note that the projectors S_i and Q_i are independent of τ_i, μ_i, whose action is only to change the norm of v_i and u_i. The vectors q_i are K-conjugate and the vectors s_i are C-conjugate.

— HB recursions of the Lanczos type: Let C be arbitrary, and τ_i and μ_i be nonzero scalars. Generate the basis vectors by the following formulas, with s_1 and q_1 arbitrary:

$$s_{i+1} = s_i - \frac{\tau_i s_i^{\mathrm{T}} C q_i}{v_i^{\mathrm{T}} A u_i} A^{\mathrm{T}} v_i \; , \tag{8.139}$$

$$q_{i+1} = q_i - \frac{\mu_i s_i^{\mathrm{T}} C q_i}{v_i^{\mathrm{T}} A u_i} A u_i \; . \tag{8.140}$$

Then (8.124) and (8.125) take the form

$$v_{i+1} = \mu_{i+1} C^{\mathrm{T}} s_{i+1} + \frac{\mu_{i+1} s_{i+1}^{\mathrm{T}} C q_{i+1}}{\mu_i s_i^{\mathrm{T}} C q_i} v_i \; , \tag{8.141}$$

$$u_{i+1} = \tau_{i+1} C q_{i+1} + \frac{\tau_{i+1} s_{i+1}^{\mathrm{T}} C q_{i+1}}{\tau_i s_i^{\mathrm{T}} C q_i} u_i \; . \tag{8.142}$$

The vectors s_i, p_i are C-conjugate. If one defines T by

$$T = S^{\mathrm{T}} A Q \; , \tag{8.143}$$

where $S = (s_1, \ldots, s_i)$, $Q = (q_1, \ldots, q_i)$, one can show that, for $C = I$, T is tridiagonal and that the scalars τ_i, μ_i can be chosen so that S and Q are the same matrices appearing in the classical algorithm of Lanczos (1950) for tridiagonalizing a matrix.

Remark 8.14
Alternative ABS formulation of the Hegedüs–Bodocs biorthogonalization recursions have been developed by Bodon (1989).

8.8 SUBCLASS S7: THE ABS CLASS OF B-CONJUGATE RESIDUALS
Subclass S7 is obtained by the following choice of the scaling vector:

$$v_i = Br_i \ , \tag{8.144}$$

where r_i is the residual, assumed to be nonzero, and B is a symmetric positive definite matrix. The following theorem establishes that the choice (8.144) is admissible and that the residuals are B-conjugate.

Theorem 8.31
If r_i is nonzero, the choice $v_i = Br_i$ is admissible and the residuals r_1, ..., r_i are B-conjugate.

Proof
For the admissibility it is enough to show that the residuals are linearly independent, which is true if r_1, ..., r_i are B-conjugate. For $i = 1$ the theorem is trivial. Assuming that the theorem is true up to $i - 1$, we observe that $r_i^T Br_i \neq 0$, since r_i is nonzero and B is positive definite, and that, from Theorem 7.5, $v_j^T r_i = r_j^T Br_i = 0$ for $j = 1, \ldots, i - 1$, establishing the B-conjugacy. Q.E.D.

Definition 8.11
The subclass of the scaled ABS algorithm where the scaling vector is given by (8.144) is called the ABS class of B-conjugate residuals.

Letting $B^i = (Br_1, \ldots, Br_i)$ the factorization relation (7.14) takes the form

$$L^i = (B^i)^T A P^i \ . \tag{8.145}$$

The following theorem indicates that L^i in (8.145) has a special structure.

Theorem 8.32
Let v_i be given by (8.144) with $r_i \neq 0$. Then the matrix L^i in (8.145) is strictly lower bidiagonal.

Proof
The elements of L^i below the diagonal have the form $(L^i)_{j,k} = r_j^T BAp_k, i \geq j > k \geq 1$. Since $r_{k+1} = r_k - \alpha_k Ap_k$ and $\alpha_k \neq 0$ from the linear independence of $r_1, \ldots, r_i$, we can write

$$(L^i)_{j,k} = \frac{r_j^T B(r_k - r_{k+1})}{\alpha_k} \; . \tag{8.146}$$

From the B-conjugacy of the residuals we get $(L^i)_{j,k} = 0$ for $j > k+1$ and $(L^i)_{k+1,k} = -r_{k+1}^T Br_{k+1}/\alpha_k \neq 0$, establishing the thesis. Q.E.D.

Some natural choices for B are the following.

— $B = I$: in this case the residuals are orthogonal.
— $B = A$, if A is square symmetric positive definite: in this case the residuals are A-conjugate.
— $B = A^{-T}$, if A is square symmetric positive definite: note that for this choice the inverse does not appear in the update formula for H_i, which reads

$$H_{i+1} = H_i - H_i r_i w_i^T H_i \; . \tag{8.147}$$

If $H_1 = I$, and z_i is proportional respectively to $A^T r_i$, $A^2 r_i$, and r_i in the three above-considered cases and w_i is proportional to z_i, then the sequence H_i consists of symmetric matrices, the update (7.9) can be written in the form (8.6) and the search vectors are orthogonal.

If z_i is scaled so that $p_i^T p_i = 1$, from (8.145) we obtain, multiplying for the transpose of both members and dropping for simplicity the index i,

$$R^T B^T A A^T B R = T \; , \tag{8.148}$$

where $T = (L^i)(L^i)^T$ is a tridiagonal matrix. If $B = I$, the columns of R are orthogonal, hence (8.148) defines an orthogonal reduction of AA^T into a tridiagonal matrix. If $B = A^{-T}$, (8.148) gives the residual relation $R^T R = T$.

8.9 SUBCLASS S8: A CLASS OF ABS UPDATES FACTORIZING A INTO LOWER HESSENBERG FORM

Subclass S8 is obtained by the following choice of the scaling vector: v_1 arbitrary and, for $i = 1, \ldots, n-1$,

$$v_{i+1} = A^T v_i \; . \tag{8.149}$$

The following theorem characterizes subclass S8.

Theorem 8.33
Suppose that the first n vectors in the Krylov sequence $(v_1, A^T v_1, \ldots)$ are linearly independent. Consider an algorithm of the scaled ABS class where v_i is the ith element of the named Krylov sequence. Then, if n iterations are required for

termination, the matrix B given by

$$B = P^{-1}AP \tag{8.150}$$

is a lower Hessenberg matrix.

Proof
If relation (8.149) is applied n times, then $v_{n+1} = A^T v_n$ is a linear combination of v_1, ..., v_n, i.e.

$$v_{n+1} = \sum_{j=1}^{n} \beta_j v_j , \tag{8.151}$$

where the β_j are some scalars. For $i = 1, \ldots, n$ the vectors on the right-hand side of (8.149) are the columns of $A^T V$. Writing the vectors on the left-hand side as the columns of a matrix VC, we have the identity

$$A^T V = VC . \tag{8.152}$$

We can determine the structure of C as follows. For $1 \leqslant i \leqslant n-1$, equating the ith column in both sides of (8.152) and taking into account (8.149) yields $v_{i+1} = Vc_i$, implying, from the linear independence of $v_1, \ldots, v_i$, that c_i is uniquely defined as $c_i = e_{i+1}$. Considering the nth column, we have $A^T v_n = Vc_n = v_{n+1}$, implying, from (8.152), that the components of c_n are $\beta_1, \ldots, \beta_n$. Hence C is a companion matrix of the following upper Hessenberg form:

$$C = \begin{bmatrix} 0 & 0 & \ldots & 0 & \beta_1 \\ 1 & 0 & \ldots & 0 & \beta_2 \\ 0 & 1 & \ldots & . & . \\ . & . & \ldots & . & . \\ . & . & \ldots & . & . \\ 0 & . & \ldots & 1 & \beta_n \end{bmatrix} \tag{8.153}$$

Define now the (nonsingular) matrix B by

$$B = L^{-1}C^T L \tag{8.154}$$

where $L = V^T AP$, see (7.15). Note that B has a lower Hessenberg form. From (8.152), $C^T = V^T AV^{-T}$. Substituting in (8.154) and using (7.15), (8.150) follows immediately. Q.E.D.
   When v_i is given by (8.149), the update of H_i reads for $i > 1$

$$H_{i+1} = H_i - H_i v_{i+1} w_i^T H_i \ . \tag{8.155}$$

If $H_1 = I$, $z_i = v_{i+1} = A^T v_i$ and w_i is proportional to z_i, update (8.155) can be written in the form (8.6) and the search vectors are orthogonal. In this case we obtain an orthogonal similarity transformation of A in a lower triangular Hessenberg form. The number of multiplications is of order $(5/2)n^3$ if the algorithm is implemented in the standard scaled Huang form. The number can be reduced to $2n^3$ or increased to $(7/2)n^3$ by using alternative implementations, as discussed in Chapter 5. The classical Householder reduction to the Hessenberg form requires of the order of $(7/3)n^3$ multiplications (Martin and Wilkinson 1968). We can also choose the available parameters as in the implicit LU factorization algorithm ($H_1 = I$, $z_i = e_i$, $w_i = e_i/e_i^T H_i v_{i+1}$, assuming that the denominator in w_i is nonzero). In such a case and disregarding pivoting operations, the number of multiplications is of the order of $(5/3)n^3$.

8.10 BIBLIOGRAPHICAL REMARKS

Almost all the results given in this chapter are due to Abaffy and Spedicato, some of them having been published previously, (Abaffy and Spedicato 1985, 1988, Abaffy 1987c, 1988d, 1988e), but many of them appearing here for the first time. Theorem 8.1 has appeared in the paper by Abaffy and Galantai (1986). Theorems 8.5, 8.8, 8.9, 8.12, 8.13, 8.15, 8.16, 8.20, 8.21, 8.22, 8.23, 8.28, 8.29, 8.30, 8.33 and the lemmas have been proved by Abaffy, and the others by Spedicato.

9

Least squares solution of overdetermined linear systems

9.1 INTRODUCTION

In this chapter we assume that A is m by n, $m \geqslant n$ and rank$(A) = q \leqslant n$, and we present several approaches for solving via the ABS algorithm the overdetermined linear system $Ax = b$ in the least squares sense. As is well known, the least squares solution x^+ of the overdetermined system $Ax = b$ is defined as any vector x^+ that minimizes the squared Euclidean norm of the residual, i.e. the function $F(x) = (Ax-b)^T(Ax-b)$. Such a vector satisfies the always compatible normal equations attributed to Gauss (1821):

$$A^T A x = A^T b. \tag{9.1}$$

If A is full rank, then $A^T A$ is positive definite and the unique solution of the normal equations can be expressed formally by

$$x^+ = (A^T A)^{-1} A^T b. \tag{9.2}$$

If A is rank deficient, one is usually interested in the particular solution of the normal equations which is of minimal Euclidean norm. Such a solution is unique and can be formally expressed in terms of A^+, the Moore–Penrose pseudoinverse of A, by

$$x^+ = A^+ b. \tag{9.3}$$

In the literature, several methods have been presented for computing the (minimal Euclidean norm) leastsquares solution of $Ax = b$. Most of the recommended methods do not form the normal equations or compute A^+ but are based upon orthogonal factorizations (of the QR or LQ type) or singular value decomposition of A or Gaussian elimination on the equivalent extended system (where r is the residual)

$$\begin{bmatrix} I & -A \\ A^T & 0 \end{bmatrix} \begin{bmatrix} r \\ x \end{bmatrix} = \begin{bmatrix} -b \\ 0 \end{bmatrix}. \tag{9.4}$$

See for a detailed analysis of these and other methods Hanson and Lawson (1974), Golub and Van Loan (1983) and Björck (1986).

In this chapter we present several approaches for finding the (minimal Euclidean norm) least squares solution in the ABS framework without directly forming and solving the normal equations. In the next chapter we present extensive numerical evidence showing that the ABS algorithms are efficient and numerically stable and that some of them provide a more accurate solution than the standard methods available in some current libraries, particularly for ill-conditioned and rank-deficient problems.

9.2 SOLUTION IN $n + m$ STEPS VIA THE HUANG ALGORITHM ON AN EXTENDED SYSTEM

We start from the observation that the normal equations are equivalent to the following two systems, of $n + m$ equations in the unknowns $x \in R^n$ and $y \in R^m$:

$$Ax = y, \tag{9.5}$$

$$A^Ty = A^Tb. \tag{9.6}$$

The above systems can be written compactly as

$$\begin{bmatrix} I_m & -A \\ A^T & 0 \end{bmatrix} \begin{bmatrix} y \\ x \end{bmatrix} = \begin{bmatrix} 0 \\ A^Tb \end{bmatrix}. \tag{9.7}$$

Note that system (9.7) has the same coefficient matrix as the system (9.4), which was originally considered by Björck (1967), on a suggestion of Golub. A generalization of system (9.7), with arbitrary right-hand side and arbitrary positive definite matrix B replacing I_m, which appears in quadratic programming, has been considered by Yang (1988a) and solved by the Huang method.

We solve (9.7) by dealing separately with (9.5) and (9.6) and by using the fundamental property of the Huang algorithm that x_{i+1} is the minimal Euclidean norm solution of the first i equations of an underdetermined system, when x_1 is a multiple of the first row of the coefficient matrix (Theorems 5.4 and 5.5). We observe that, for the overdetermined system (9.5) to be solvable, y must lie in the range of A or equivalently in the row space of A^T. Now it is known (see for instance Albert (1972)) that there exists a unique solution y^+ of an underdetermined system which lies in the row space of the coefficient matrix and that solution is the (unique) minimal Euclidean norm solution. Therefore, (9.5) is solvable if and only if y is the minimal Euclidean norm solution y^+ of the underdetermined system (9.6). From

Theorems 5.4 and 5.5 this is the case when (9.6) is solved by the Huang algorithm with x_1 proportional to the first column of A. Once y^+ is obtained, then any algorithm of the ABS class can be used to solve (9.5) for the unique solution if A is full rank ($m - n$ linearly dependent equations will be removed by the simultaneous verification of the equalities $s_i = 0$ and $\tau_i = 0$ in step (C1) of ALGORITHM 1). If A is rank deficient and the minimal Euclidean norm solution is wanted, then it is enough to use again the Huang algorithm with x_1 a multiple of the first row of A ($m - q$ equations will be now removed in step (C1)).

9.3   SOLUTION IN *n* STEPS VIA A SUBCLASS OF THE SCALED ABS ALGORITHM

When m is greater than n, we can still apply the scaled ABS algorithm to the system for n steps corresponding to n linearly independent equations, provided that the corresponding scaling vectors are linearly independent. Suppose now that A is full rank and consider the subclass of the scaled ABS class where v_i is given by

$$v_i = Au_i , \tag{9.8}$$

where $u_i \in R^n$ is the ith column of a nonsingular matrix $U = (u_1, \ldots, u_n)$. The vectors v_i are linearly independent and therefore admissible. After n steps the residual r_{n+1} satisfies, from Theorem 7.5, the relations

$$u_j^T A^T r_{n+1} = 0 , \quad j = 1, \ldots, n, \tag{9.9}$$

or in matrix form

$$U^T A^T r_{n+1} = 0. \tag{9.10}$$

Since U is nonsingular and $r_{n+1} = Ax_{n+1} - b$, (9.10) implies that x_{n+1} satisfies the normal equations and thus it is the required least squares solution.

If $u_i = e_i$, v_i is the ith column of A and we get algorithms for least squares of the type considered by Stewart (1973). If $u_i = p_i$, the resulting algorithms coincide with subclass S3, the orthogonally scaled ABS class, which contains the implicit QR algorithm. We state this result as the following.

Theorem 9.1
If the implicit QR algorithm is applied to the overdetermined linear system $Ax = b$, where A is full rank, then it terminates at the nth step having computed the least squares solution.

The recursions of the scaled ABS algorithm where v_i is given by (9.8) can be considered either as the recursions of the standard ABS algorithm applied on the scaled system $U^T A^T Ax = U^T A^T b$ or as the recursions of the scaled ABS algorithm with $v_i = u_i$ applied on the normal system, implying that definition (9.8) is admissible even if A is rank deficient. Since the normal system is compatible, it follows that, in the rank-deficient case, both s_i and τ_i are zero in step (C5) of ALGORITHM 5 whenever a linearly dependent equation of the normal system is detected. Thus the algorithms discard such an equation and x_{n+1} still solves the normal system. In order that x_{n+1} be the minimal norm solution in such a case, it is enough, from Theorem 8.3, to use the scaled Huang algorithm of the considered subclass (i.e. $H_1 = I$, z_i a multiple of $A^T A u_i$, H_i updated by (8.6)) and x_1 proportional to $A^T A u_1$). Note that, while the choice $u_i = e_i$ is feasible, the choice $u_i = p_i$ is not computionally acceptable, since z_i would be an eigenvector of $H_i A^T A$ (and p_i an eigenvector of $A^T A$). If the implicit QR algorithm is used on a rank-deficient problem, with $x_1 = 0$, then x_{n+1} is not a minimal norm solution but a basic-type solution (where the last $n - q$ components are equal to zero).

9.4 SOLUTION BY EXPLICIT CONSTRUCTION OF ORTHOGONAL FACTORIZATIONS

Another approach to the least squares solution via the ABS algorithm is obtained by explicit construction of the underlying factorization of A, which is then used in the traditional way. Here we consider the orthogonal factorizations associated with the Huang and the implicit QR algorithm.

Let A be full rank and construct normalized search vectors $p_1, \ldots, p_n \in R^m$ by applying the Huang algorithm (with $z_i = a_i / \|H_i a_i\|$) to the matrix A^T. From the underlying factorization $A^T P = L$, we get

$$P^T A = R, \tag{9.11}$$

with R nonsingular upper triangular. To show that (9.11) implies that $A = PR$, suppose that $m - n$ linearly independent columns are added to A, so that the factorization reads

$$P'^T A' = R', \tag{9.12}$$

where $A' = (A, \overline{A})$, $P' = (P, \overline{P})$ and

$$R' = \begin{bmatrix} R & S \\ 0 & \overline{R} \end{bmatrix},$$

$\overline{A}$ consisting of the added columns, $\overline{R}$ being nonsingular upper triangular and P' being now orthogonal. Premultiplying (9.12) by P' gives $A' = P'R'$ or

$$(A, \overline{A}) = (PR, PS + \overline{P}\overline{R}). \tag{9.13}$$

Hence

$$A = PR. \tag{9.14}$$

Using (9.14) in the normal equations, we get $R^TP^TPRx = R^TP^Tb$ or, since $P^TP = I_n$ and R is nonsingular,

$$Rx = P^Tb. \tag{9.15}$$

Therefore the unique least squares solution is obtained by solving (9.15) by back substitution. The outlined procedure is essentially the standard Gram–Schmidt procedure for linear least squares, differing in the way that the factors are computed.

The factorization (9.14) can also be obtained by applying the implicit QR algorithm to the matrix A for n steps. In this case the underlying factorization becomes

$$V^TA = DR^{-1} , \tag{9.16}$$

where $R = P$ and V^T is n by m with orthogonal rows. Proceeding as before, one can show that (9.16) implies that $A = VDR^{-1}$. Substituting in the normal equations and using the relation $V^TV = I_n$, assuming that the v_i are normalized, we obtain

$$DR^{-1}x = V^Tb. \tag{9.17}$$

Hence the least squares solution is directly expressed by

$$x^+ = PD^{-1}V^Tb. \tag{9.18}$$

If A is rank deficient and we are using the factorization associated with the Huang algorithm, we can proceed as follows. Whenever linear dependence of a column of A is detected by the occurrence $s_i = 0$, let us interchange such a column with one of the following ones which is not dependent. This results in putting the $n - q$ dependent columns in the last $n - q$ positions. Letting Q be the permutation matrix that implements such interchanges, it is clear that the following factorization is finally obtained:

$$AQ = PR , \tag{9.19}$$

where P is m by q full rank with orthonormal columns and R is q by n full rank, the matrix comprising the first q columns of R being moreover nonsingular upper triangular. Using (9.19) in the normal equations yields

$$RQ^Tx = P^Tb \tag{9.20}$$

which is an underdetermined full rank system. Thus the minimal Euclidean norm solution can be obtained using the Huang algorithm with x_1 a multiple of the first row of RQ^T (for instance $x_1 = 0$).

9.5   SOLUTION VIA EXPLICIT EVALUATION OF THE MOORE–PENROSE PSEUDOINVERSE

As observed before, the minimal norm least squares solution of $Ax = b$ can be expressed by (9.3) in terms of the Moore–Penrose pseudoinverse A^+. In Chapter 5 some connections between the Huang algorithm and the Moore–Penrose pseudo-inverse have been considered. Among them we gave a modification of the Pyle method for computing A^+. Such a method can be used both in the full rank and in the rank-deficient case. Here we consider in more detail an alternative representation of the pseudoinverse in terms of the modified Huang algorithm. Let us first assume that A is full rank. Let us denote A^T by $A' = (a_1', \ldots, a_n')^T$, $a_i' \in R^m$. We consider the representation of A'^+ from which A^+ follows through the equality $A^+ = (A'^+)^T$. Using (4.6) with $H_1 = I$, (5.27) and the identity of the matrices generated by the Huang and the modified Huang algorithm we have, with W^i the matrix of the parameters w_i appearing in the modified Huang algorithm and $G_i = (A'^{iT}A'^i)^{-1}$,

$$G^i W^{iT} = (A'^{iT}A'^i)^{-1}A'^{iT} \tag{9.21}$$

or, from (5.62),

$$A'^{i+} = W^i G^{iT}. \tag{9.22}$$

Using Theorem 4.7 and the definition $w_i = H_i a_i'/(H_i a_i')^T(H_i a_i')$ it is easily seen by induction that G^i is an upper triangular matrix with units on the diagonal and that it can be recursively computed through the relation

$$G^i = \begin{bmatrix} G^{i-1} & g_i \\ 0 & 1 \end{bmatrix}, \tag{9.23}$$

with $G^1 = 1$ and

$$g_i = G^{i-1}(W^{i-1})^T a_i'. \tag{9.24}$$

The vector w_i can be expressed through the following formulas (see (5.33) with $p_i = H_i a_i'$):

$$s_i = a_i' - (A')^{i-1} g_i \tag{9.25}$$

$$w_i = \frac{s_i}{s_i^T s_i}. \tag{9.26}$$

It is possible to consider a further reprojection in the definition of w_i, which has turned out in the numerical experiments to give a further increase in accuracy, i.e. $w_i = H_i^2 a_i'/(H_i^2 a_i')^T (H_i^2 a_i')$. Then the formula corresponding to (9.26) is

$$w_i = \frac{s_i - (A')^{i-1} G^{i-1} (W^{i-1})^T s_i}{s_i^T s_i}. \tag{9.27}$$

After n steps the identity $A'^+ = (A'^n)^+ = (A^T)^+ = W^n (G^n)^T$ is obtained. Therefore the least squares solution is expressed by

$$x^+ = G^n (W^n)^T b. \tag{9.28}$$

Comparing (9.28) with (9.15) and observing that $W^n = PD$, D diagonal with $D_{i,i} = 1/s_i^T s_i$, we see that the difference between the pseudoinverse approach and the factorization approach lies in the fact that in the former the inverse of R is directly constructed.

If A is rank deficient, by row interchanges we can remove to the last $n - q$ positions the $n - q$ dependent rows of A'. After n steps we obtain the pseudoinverse $(\overline{A}^q)^+ = G^q (W^q)^T$ of the matrix $\overline{A}^q$ comprising the first q columns of $\overline{A} = (QA')^T$, Q being the permutation matrix used in the row interchanges in A'. Now define the matrix $C \in R^{q,m}$ and the vector $d \in R^q$ by

$$C = [I_q, (\overline{A}^q)^+ A''], \tag{9.29}$$

$$d = (\overline{A}^q)^+ b \tag{9.30}$$

where A'' consists of the last $n - q$ columns of $(QA')^T$. Consider the system

$$Cx = d. \tag{9.31}$$

Then one can show (Spedicato and Bodon 1989a) that system (9.31) is equivalent to system (9.20). Therefore the minimal Euclidean norm solution can again be determined by solving (9.31) with the Huang (or the modified Huang) method and a starting point multiple of the first row of C.

Another approach when A is rank deficient is based upon a formula due to Cline (1964), which relates the pseudoinverses of $(A')^{i-1}$ and A'^i. Define $d_i \in R^{i-1}$ by

$$d_i = [(A')^{i-1}]^+ a_i' \tag{9.32}$$

and $c_i \in R^n$ by

$$c_i = a_i' - (A')^{i-1} d_i. \tag{9.33}$$

Note that, from (5.63), we have the identity $c_i = H_i a_i' = p_i$, where H_i and p_i are generated by the Huang (or the modified Huang) method on the matrix A', and that c_i is zero if and only if a_i' depends linearly on $a_1', \ldots, a_{i-1}'$. Then the Cline formula is the following:

$$(A'^i)^+ = \{[(A')^{i-1}]^+ - y_i d_i^T, y_i\} , \qquad (9.34)$$

where, if $c_i \neq 0$,

$$y_i = \frac{c_i}{c_i^T c_i} ; \qquad (9.35)$$

otherwise

$$y_i = \frac{[(A')^{i-1}]^+ d_i}{1 + d_i^T d_i} . \qquad (9.36)$$

Thus the Cline formula can be used in the context of the previously described procedure whenever linear dependence of a row of A' is detected, y_i being then given by (9.36). It can also be used on determined or underdetermined incompatible problems to compute the minimal norm least squares solution. See for further discussion Zhao (1981).

9.6 BIBLIOGRAPHICAL REMARKS

The results presented in this chapter are almost all due to Spedicato and Bodon. The approach via the extended system (9.7) was presented by Spedicato (1985). The approach via the n-step iteration, including the QR and the Stewart methods, has appeared in the paper by Spedicato and Bodon (1987). The general scaling parameter choice (9.8) is due to Spedicato, but Bodon (1987) first observed that the implicit QR algorithm terminates on the least squares solution. The approach via pseudoinverse has been studied by Spedicato and Bodon (1989a). Further analysis with numerical experiments is available in the Ph.D. dissertation of Bodon (1987). Zhao (1981) considered connections between the Huang method and the pseudoinverse and made use of the Cline formula.

10

Computational performance of the modified Huang algorithm on linear systems and linear least squares

10.1 INTRODUCTION

In this chapter we present numerical results on the performance of the Huang algorithm and the modified Huang algorithm on linear systems and linear least squares. The Huang algorithm, with several of its alternative formulations, has been extensively tested for linear systems by Abaffy and Spedicato (1983a), Abaffy *et al.* (1983), Abaffy and Spedicato (1987), Bertocchi and Spedicato (1988a,b,c), Spedicato and Vespucci (1989) and Spedicato and Bodon (1989b). Numerical testing of the modified Huang algorithm, in the context of various ABS approaches to linear least squares, has been done by Bodon (1987) and Spedicato and Bodon (1989a, 1989c). Testing of other ABS algorithms for linear systems, particularly the implicit LU and QR algorithms, is available in the papers by Oprandi (1987), Abaffy (1987b) and Spedicato and Bodon (1989b).

The principal aim in the experiments in the quoted papers has been the following.

— The performance of alternative formulations of the algorithms has been evaluated. Abaffy and Spedicato (1987) have considered about 60 versions of the Huang algorithm, and Spedicato and Bodon (1989b) about 200 versions of the implicit LU, LQ and QR algorithms.
— The performance of a given algorithm on families of problems with growing dimensions, growing condition numbers and whose solution is exactly known has been evaluated.
— The performance of the tested ABS algorithms has been compared with that of commercially available codes implementing sophisticated versions of classical algorithms. The codes implementing the QR algorithm with Householder rotations in the NAG and LINPACK packages and the code implementing the LU factorization algorithm in the IBM ESSL library have been considered.

In view of the very large number of available numerical results and of the need for still further testing, we consider only some results from testing several versions of the Huang algorithm on linear systems and linear least squares. The results presented indicate that some implementations of the modified Huang algorithm are competitive (and usually more accurate on very-ill-conditioned problems) with the commercially available implementations of the classical algorithms. We do not claim here that the modified Huang algorithm is the best choice in the ABS class. Evidence given by Bodon (1988) and Spedicato and Bodon (1989b) shows that some implementations of the implicit LU algorithm are competitive with the modified Huang algorithm and that some implementations of the Huang algorithm via optimal conditioning or of the implicit QR algorithm with reprojections in the computation of the v_i and p_i perform even better than the modified Huang algorithm.

Finally we note that preliminary experiments of Bertocchi and Spedicato (1988b,c) on vector implementations of the implicit LU and the modified Huang algorithms on the IBM 3090 indicate that in some ranges of dimension the speed attainable by the ABS algorithm (in megaflops per second) in a FORTRAN implementation is higher than the speed reached by an Assembler implementation of the LU factorization algorithm in the ESSL library.

10.2 PERFORMANCE OF THE HUANG AND THE MODIFIED HUANG ALGORITHMS ON LINEAR SYSTEMS

In Tables 10.1–10.5 we compare the performance of eight implementations of the Huang algorithm and the modified Huang algorithm versus five classical implementations of the QR and the Gram–Schmidt algorithms. The tested algorithms correspond to the following symbols.

HS: The standard Huang algorithm, where
$p_i = H_i a_i, \alpha_i = \tau_i / p_i^T a_i$, where $\tau_i = a_i^T x_i - b^T e_i$, and $H_{i+1} = H_i - p_i p_i^T / p_i^T a_i$.

MHS: The modified Huang algorithm, where
$z_i = H_i a_i, p_i = H_i z_i, \alpha_i = \tau_i / z_i^T z_i$, and $H_{i+1} = H_i - z_i p_i^T / z_i^T z_i$.

MHST: The modified Huang algorithm, differing from MHS in using $\alpha_i = \tau_i / p_i^T p_i$ and $H_{i+1} = H_i - p_i p_i^T / p_i^T p_i$.

HBE: The formulation of the standard Huang algorithm, where p_i is computed by (5.30)–(5.33).

MHBE: The formulation of the modified Huang algorithm, where p_i is computed by (5.30)–(5.34).

HB: The formulation of the standard Huang algorithm where p_i is computed by (5.37).

MHB: The formulation of the modified Huang algorithm, where p_i is computed by (5.41) and (5.42).

HCS: The formulation of the standard Huang algorithm where p_i is computed by ALGORITHM 3 (with $u_j^1 = a_j, j = 1, \ldots, n$).

GIVENS: The algorithm based upon explicit QR factorization built using Givens rotations (code produced at N.O.C., Hatfield Polytechnic).

HOUSE: The algorithm based upon explicit QR factorization built using Householder rotations (code from the LINPACK library).

BRENT: The algorithm based upon the Brent method for nonlinear systems, (see
section 5.6); code produced by More and Cosnard (1979).

DANIEL: The algorithm based upon the reorthogonalized Gram–Schmidt procedure of Daniel et al. (1976), i.e. (5.43) and (5.44).

MODGS: The algorithm using the modified Gram–Schmidt procedure, i.e. (5.45).

Tables 10.1–10.4 have been obtained by testing the algorithms on 33 ill-
conditioned problems of dimension between $n = 4$ and $n = 20$ and condition number
between 4×10^3 and 8×10^{13} (see Spedicato and Vespucci (1989) for the precise
definition of the problems). General features of the problems are the following.

— The elements of A are integers and A is nonsingular.
— A solution x^+ is assigned with integer components randomly defined in the
interval $[-50, 50]$.
— The right-hand side b is defined by computing Ax^+, with a check that no integer
overflow occurs.
— For each matrix A, 100 problems are defined by computing 100 right-hand sides.
— The computer is an IBM 4361 in single precision (machine zero, about 10^{-7}).

In the tables the symbols heading the columns have the following meaning.

ESMIN: Minimum value of the relative error in the solution $\sigma^+ = \|x_{n+1} - x^+\|/\|x^+\|$ (with respect to the tested right-hand sides).
ESMAX: Maximum value of σ^+.
ESMED: Geometric average value of σ^+.
ERMED: Geometric average value of the relative residual error $\tau^+ = \|Ax_{n+1} - b\|/\|b\|$ (with respect to the tested right-hand sides); note that the
residuals have been computed in double precision.

Note that, in the paper by Spedicato and Vespucci (1989), additional error
measures are tabulated.

In Table 10.1 the geometric averages are given, with respect to the 33 tested
problems, of the previously defined errors. Algorithms HS, HCS, HB perform
similarly, while the performance of HBE is poorer. Algorithms MHS, MHB, MHST
also perform similarly and one order better than HS, HCS, HB, while MHBE
performs poorer. Among the algorithms inplementing classical procedures, HOUSE
DANIEL and BRENT perform very similarly and are the best. Their performance is
similar to that of the modified Huang algorithm in σ^+ but is one order better in the
residual error.

Table 10.2 gives the maximum error in all problems, which is an indicator of the
robustness of the algorithms. Here the best performance is that of the algorithms
DANIEL, MHS, in terms of σ^+, while HOUSE and DANIEL give the best result in
terms of τ^+.

Table 10.3 gives results on the ill-conditioned Micchelli problem defined by $A_{i,j} = |\alpha_i, \alpha_j|$, where the α_j are distinct integers. We can draw the same conclusions as from
Table 10.1.

Table 10.4 gives the result on a very-ill-conditioned problem using the Pascal

Table 10.1 — Geometric averages on all problems

	ESMIN	ESMAX	ESMED	ERMED
HS	1E − 2	7E − 1	2E − 1	4E − 5
HCS	7E − 3	7E − 1	2E − 1	2E − 5
MHS	1E − 3	1E − 1	3E − 2	6E − 5
HBE	5E − 2	1E0	5E − 1	2E − 3
MHBE	3E − 2	1E0	5E − 1	6E − 3
HB	1E − 2	7E − 1	3E − 1	1E − 4
MHB	2E − 3	5E − 1	8E − 2	7E − 6
MHST	2E − 3	5E − 1	8E − 2	4E − 6
GIVENS	2E − 3	4E0	6E − 1	7E − 4
HOUSE	1E − 3	4E − 1	8E − 2	2E − 6
DANIEL	1E − 3	1E − 1	3E − 2	1E − 6
MODGS	2E–2	6E − 1	3E − 1	3E − 4
BRENT	2E − 3	4E − 1	8E − 2	1E − 6

Table 10.2 — Maximum errors on all problems

	ESMIN	ESMAX	ESMED	ERMED
HE	5E − 1	5E0	1E0	9E − 4
HCS	5E − 1	2E + 1	4E0	8E − 3
MHS	2E − 1	4E0	9E − 1	4E − 1
HBE	6E − 1	3E0	1E0	7E − 2
MHBE	7E − 1	2E + 2	2E + 1	5E0
HB	5E − 1	1E + 1	5E0	1E − 2
MHB	7E − 1	7E + 2	7E + 1	3E − 2
MHST	2E0	5E + 2	1E + 2	2E − 1
GIVENS	1E0	Overflow	Overflow	Overflow
HOUSE	3E0	3E + 2	4E + 1	5E − 6
DANIEL	2E − 1	2E − 0	7E − 1	7E − 6
MODGS	5E − 1	7E0	2E0	1E − 2
BRENT	7E − 1	1E + 2	2E + 1	1E − 5

matrix, which is defined by $A_{i,j} = A_{i-1,j} - A_{i,j-1}$ for $i,j = 2, \ldots, n$, $A_{1,j} = A_{j,1} = 1$, $j = 1, \ldots, n$. The best performance in terms of σ^+ is shown by MHS and DANIEL. Among the classical procedures, HOUSE and BRENT give a smaller error in τ^+ than MHS does, but their computed solution has generally not a single correct digit.

Table 10.3 — The Michelli problem: $n = 20$, Cond $= 2E4$

	ESMIN	ESMAX	ESMED	ERMED
HS	7E − 3	3E − 1	8E − 2	2E − 4
HCS	4E − 3	1E − 1	4E − 2	3E − 4
MHS	1E − 4	4E − 3	9E − 4	2E − 6
HBE	6E − 1	1E0	8E − 1	7E − 2
MHBE	4E − 1	9E0	2E0	9E − 2
HB	5E − 2	4E − 1	2E − 1	1E − 3
MHB	1E − 4	3E − 3	8E − 4	2E − 6
MHST	1E − 4	4E − 3	9E − 4	2E − 6
GIVENS	7E − 5	5E − 3	1E − 3	5E − 6
HOUSE	9E − 5	9E − 4	3E − 4	3E − 6
DANIEL	1F − 4	3F − 3	8F − 4	3E − 6
MODGS	2E − 3	2E − 1	8E − 2	3E − 4
BRENT	1E − 4	5E − 3	1E − 3	3E − 6

Table 10.4 — The Pascal problem $n = 11$, Cond $= 1E11$

	ESMIN	ESMAX	ESMED	ERMED
HS	2E − 1	9E − 1	6E − 1	1E − 5
HCS	5E − 1	2E + 1	4E0	2E − 5
MHS	1E − 1	9E − 1	5E − 1	8E − 2
HBE	3E − 1	1E0	8E − 1	1E − 2
MHBE	2E − 1	1E0	7E − 1	4E − 4
HB	2E − 1	9E − 1	6E − 1	2E − 4
MHB	6E − 1	2E + 2	5E + 1	2E − 3
MHST	5E − 1	6E + 1	1E + 1	2E − 4
GIVENS	4E − 1	2E + 1	4E0	3E − 5
HOUSE	3E0	3E + 2	4E + 1	2E − 6
DANIEL	1E − 1	2E0	5E − 1	2E − 6
MODGS	2E − 1	1E0	7E − 1	2F − 5
BRENT	2E − 1	1E + 2	2E + 1	4E − 6

Table 10.5 is taken from experiments by Bertocchi and Spedicato (1988a) on an IBM 3090 in double precision (about 15 decimal digits; machine zero, 2^{-64}). The performance of MHST is compared with that of the *LU* code in the IBM ESSL library on a Pascal problem of dimension varying up to $n = 17$ and with 1000

Table 10.5 — Performance of the MHST and *LU* codes in the ESSL library

Method	n	COND	ESMIN	ESMAX	ESMED	ERMED
ESSL	7	2E6	5E − 14	3E − 10	4E − 12	1E − 16
MHST	7		8E − 14	2E − 11	4E − 12	1E − 16
ESSL	9	5E8	7E − 13	5E − 10	6E − 11	1E − 16
MHST	9		2E − 12	4E − 9	4E − 10	1E − 16
ESSL	11	1E11	1E − 9	1E − 6	1E − 7	2E − 16
MHST	11		1E − 10	2E − 7	3E − 8	2E − 16
ESSL	13	2E13	1E − 6	4E − 4	9E − 5	2E − 16
MHST	13		7E − 9	2E − 5	2E − 6	2E − 16
ESSL	15	4E15	3E − 5	6E − 2	8E − 3	3E − 16
MHST	15		1E − 6	3E − 2	3E − 4	2E − 16
ESSL	16	6E16	2E − 4	2E0	9E − 2	2E − 16
MHST	16		2E − 5	2E − 2	3E − 3	1E − 16
ESSL	17	9E17	7E − 4	3E0	1E − 1	2E − 16
MHST	17		9E − 5	2E − 1	3E − 2	1E − 16

randomly generated integer right-hand sides (the estimated condition numbers are provided). From the table it appears that MHST for $n > 10$ gives smaller σ^+, while τ^+ is of similar size. Note that τ^+ almost does not change with the increase in the condition number, which affects evidently σ^+.

From the presented tables and drawing also on the additional results available in the quoted papers, we deem that the following conclusions are justified

— The various versions of the Huang algorithm perform similarly, with the exception of HBE, whose performance is inferior.
— The various versions of the modified Huang algorithm perform similarly, with the exception of MHBE, whose performance is inferior.
— The modified Huang algorithm performs generally more accurately than the standard Huang algorithm and similarly with the algorithm of Daniel *et al*.
— The modified Huang algorithm is competitive with the best considered commercially available codes. On ill-conditioned problems it tends to give larger errors in the residual, but it usually gives smaller errors in the approximation x_{n+1} of the solution.

10.3 PERFORMANCE OF THE MODIFIED HUANG ALGORITHM ON LINEAR LEAST SQUARES

Extensive numerical experiments on several ABS algorithms for linear least squares have been made by Bodon (1987) and Spedicato and Bodon (1989a,c). Here we give only results corresponding to some of the tested algorithms, which are indicated in the tables by the following symbols.

EXT: The algorithm where systems (9.6) and (9.5) are solved for the minimal
            Euclidean norm solution using the modified Huang algorithm with the
            null vector as starting point. For the solution of (9.6) the modified Huang
            method is implemented using (5.41) and (5.42), which lead to a saving in
            the number of multiplications (always) and in the storage requirement (if
            $m > n/2$) with respect to the standard implementation via (5.21) and
            (5.24).

FACT: The algorithm based upon the explicit factorization (9.14), where P and
            R are computed via the modified Huang algorithm using (5.41) and
            (5.42).

QR: The algorithm based upon the direct application on the overdetermined
            system $Ax = b$ of the implicit QR algorithm. The considered formulation
            of the QR algorithm is the one where $z_i = e_i$, $v_i = Ap_i$, $H_1 = I$ and H_i is
            updated by (8.63), requiring of the order of $(3/2)mn^2$ multiplications and
            $n^2/4$ storage locations.

HOUSE: The code of the NAG library implementing the classical QR factorization
            algorithm with Householder notations, modified to eliminate column
            pivoting when experimenting with full rank problems.

The codes have been run in single precision on a Burroughs 6800 computer
(machine zero, about 10^{-12}). For the ABS algorithm the dependency test $H_iA^Tv_i = 0$ has been implemented as the test $\|H_iA^Tv_i\| < 10^{-6}\|A^Tv_i\|$.

The Tables refer to the testing over a total of 405 problems, of which 243 are
considered well-conditioned and 45 ill-conditioned and 117 have rank-deficient
matrices. The well-conditioned and the ill-conditioned problems belong to a family
of test problems, introduced by Spedicato and Bodon (1987), where the entries of the
problem are exactly represented in the machine and the solution is exactly known.
For the full rank case the family is defined as follows. We observe from (9.2) that, if
x^+ is a least squares solution of the overdetermined system $Ax = b$, then it is also a
solution of $Ax = b + \bar{b}$, where $\bar{b}$ is orthogonal to the columns of A. Let $\bar{b} \in R^m$ be an
arbitrary vector with integer components and the first component equal to -1.
Partition $\bar{b}$ and A in the form

$$\bar{b}^T = (-1, b_1^T, b_2^T), \quad b_1 \in R^{m-n-1}, \quad b_2 \in R^n, \tag{10.1}$$

$$A^T = (c, S^T, R^T) \tag{10.2}$$

where $c \in R^n$, $S = (s_1, \ldots, s_n)$, $s_i \in R^{m-n-1}$, $R = (r_1, \ldots, r_n)$, $r_i \in R^n$. Let the
elements of S and R be arbitrary integers, with the condition that R is nonsingular
(Spedicato and Bodon (1987) took R to be upper triangular with nonzero diagonal
elements). Condition $A^T\bar{b} = 0$ can be satisfied by defining the integer elements of c
componentwise as follows:

$$c_j = s_j^T b_1 + r_j^T b_2, \quad j = 1, \ldots, n. \tag{10.3}$$

Let x^+ be an arbitrary vector with integer entries and compute the integer vector

$b = Ax^+$. Then the unique least squares solution of $Ax = b + \bar{b}$ is x^+ and the final residual is $-\bar{b}$.

By relinquishing the property that the entries be integers but still allowing exact machine representation, we can modify the above-defined family so that the problems are full rank and ill-conditioned. Define as before the first $n-1$ columns of S and R. Define the last column of S and R as a linear combination with integer coefficients of the first $n-1$ columns, save the element $R_{n,n}$, which is set equal to 2^{-k}, k a preassigned integer, such that 2^{-k} is greater than the machine zero. Define $\bar{b}$ as before, save that the last component is zero. Then the last column of A is a linear combination of the previous columns, save for the element $A_{n,n}$. A is full rank, but getting closer to a matrix of rank $n-1$ as k increases. Define the components of x^+ as arbitrary integers save that the last component is equal to 2^k. Then no round-off occurs in the definition of the entries and the condition number is expected to increase with increasing k.

Particular problems in this family have been obtained by relations of the form $(b_1)_j = [K_1 q]$, $(b_2)_j = [K_1 q]$, $A_{i,j} = [K_2 q]$, $x_j^+ = [K_3 q]$, where K_1, K_2, K_3 are given integers, q is a uniformly distributed random number in $[-1, +1]$. The 243 'well-conditioned' problems have been generated by the 243 combinations of the values $n = 5, 10, 20$, $m = n + 1, 2n, 5n$ and $K_1, K_2, K_3 = 10, 100, 1000$. The 45 'ill-conditioned' problems have been obtained by setting $K_1 = K_2 = K_3 = 10$, $k = 4, 6, 8, 10, 12$, n and m as before.

The rank-deficient problems also have integer entries for A, b and x^+; they belong to the following family introduced by Zielke (1986). Let $D \in R^{q,q}$ be a diagonal matrix whose diagonal elements are negative powers of 2 and are greater than the machine zero. Let $u \in R^m$, $v \in R^n$, $m > n > q$, be vectors with arbitrary integer components. Define the matrix $B \in R^{m,n}$ through the following nonsingular value decomposition:

$$B = \left(\frac{I_m - 2uu^T}{u^T u} \right) \begin{bmatrix} D & 0 \\ 0 & 0 \end{bmatrix} \left(\frac{I_n - 2vv^T}{v^T v} \right). \tag{10.4}$$

Define the matrix $A \in R^{m,n}$ by

$$A = (u^T u v^T v)B. \tag{10.5}$$

Then A is integer valued of rank q, its nonsingular values are those of B multiplied by $u^T u v^T v$ and its condition number is the same as B. The pseudoinverse A^+ of A has the form

$$A^+ = (v^T v I_n - 2vv^T) \begin{bmatrix} D^{-1} & 0 \\ 0 & 0 \end{bmatrix} \frac{v^T v I_m - 2uu^T}{(u^T u v^T v)^2}. \tag{10.6}$$

Thus, given a vector $b \in R^m$ whose components are integer multiples of $(u^T u v^T v)^2$, the vector $x^+ = A^+ b$ has integer components and is the minimal Euclidean norm least squares solution of $Ax = b$.

Table 10.6 — Relative error ES in 243 well-conditioned problems

	EXT	QR	FACT	HOUSE
EXT		104/4	5	5/1
QR	135/4		2	2
FACT	238	241		163/6
HOUSE	237/1	240	74/6	

Score: FACT (642/6), HOUSE (551/7), QR (139/4), EXT (114/6).

Table 10.7 — Relative error ER in 243 well-conditioned problems

	EXT	QR	FACT	HOUSE
EXT		99/6	63/5	129/6
QR	138/6		75/4	147/3
FACT	173/5	164/4		180/5
HOUSE	108/6	93/4	58/5	

Score: FACT (519/14), QR (360/13), EXT (291/17), HOUSE (259/15).

Table 10.8 — Relative error ES in 45 ill-conditioned problems

	EXT	FACT	HOUSE
EXT		9	9
FACT	36		28/1
HOUSE	36	16/1	

Score: FACT (64/1), HOUSE (54/1), EXT (18).

In Tables 10.6–10.11 we give numbers which indicate how many times the algorithm at the row heading is better than the algorithm at the column heading (the second number indicates how many times the first two digits are the same). The best performance is shown by the ABS algorithm FACT. The superiority with respect to

Table 10.9 — Relative errors ER in 45 ill-conditioned problems

	EXT	FACT	HOUSE
EXT		8/2	24/1
FACT	35/2		39
HOUSE	20/1	6	

Score: FACT (74/2), EXT (32/3), HOUSE (26/1).

Table 10.10 — Relative error ES in 117 rank-deficient problems

	EXT	FACT	HOUSE
EXT		4	16/1
FACT	113		80
HOUSE	100/1	37	

Score: FACT (193), HOUSE (137/1), EXT (20/1).

Table 10.11 — Relative error ER in 117 rank-deficient problems

	EXT	FACT	HOUSE
EXT		65/2	83
FACT	50/2		102/1
HOUSE	34	14/1	

Score: FACT (152/3), EXT (145/2), HOUSE (48/1).

the NAG code is particularly evident in the ill-conditioned and rank-deficient problems. The superiority is also more remarkable with respect to the error ER in the residual than to the error ES in the solution, in contrast with what is observed for linear systems. Finally in Table 10.12 we present results on ill-conditioned problems by algorithm FACT, where three different formulas have been used for building the vectors p_i; SGS corresponds to using the stabilized Gram–Schmidt formula (5.45), SGS2 corresponds to applying (5.45) twice, in the spirit of the Daniel *et al.* iterated application of the unstabilized Gram–Schmidt procedure, while MHUA corre-

Table 10.12 — Comparison of the stabilized Gram–Schmidt, the iterated stabilized Gram–Schmidt and the modified Huang methods

m	n	k	Method	ES	ER
6	5	4	SGS	$2E-10$	$2E-16$
6	5	4	MGS2	$2E-15$	$8E-17$
6	5	4	MHUA	$9E-14$	$6E-17$
6	5	12	SGS	$9E-6$	$3E-16$
6	5	12	MGS2	$2E-12$	$2E-16$
6	5	12	MHUA	$3E-11$	$7E-17$
10	5	4	SGS	$7E-11$	$9E-17$
10	5	4	MGS2	$1E-13$	$1E-16$
10	5	4	MHUA	$3E-14$	$3E-17$
10	5	12	SGS	$3E-5$	$1E-16$
10	5	12	MGS2	$4E-11$	$1E-16$
10	5	12	MHUA	$9E-12$	$1E-18$
11	10	4	SGS	$2E-10$	$1E-16$
11	10	4	MGS2	$2E-13$	$1E-17$
11	10	4	MHUA	$7E-14$	$1E-16$
11	10	12	SGS	$1E-5$	$2E-16$
11	10	12	MGS2	$9E-11$	$4E-16$
11	10	12	MHUA	$3E-11$	$2E-16$

sponds to the use of (5.41) and (5.42), as was the case in Tables 10.6–10.11. From the table it appears that there is little difference in the residual error and that MHUA and MGS2 perform equivalently, while SGS is many orders less accurate.

10.4 BIBLIOGRAPHICAL REMARKS

The results presented in section 10.2 are from Spedicato and Vespucci (1989), and those in section 10.3 from Spedicato and Bodon (1989c).

11

Application of the implicit *LU*, *LQ* and *QR* algorithms to some sparse large linear systems

11.1 INTRODUCTION

In this chapter we consider the application of the implicit LU, LQ and QR algorithms to some types of sparse and possibly very large linear system. The sparse matrices that we consider are the following.

— Positive definite matrices having the m-block nested dissection form (later referred to as ND matrices), which arise in particular when the nested dissection ordering technique of George (1973) is applied to matrices obtained from regular finite element meshes
— Standard q-banded matrices, upper q-banded matrices and modified q/p banded or upper q/p-banded matrices.

With regard to ND matrices we show that the implicit LU algorithm generates matrices H_i where the nonzero elements are found only in positions corresponding to the nonzero blocks of the ND matrix. From this fact it follows that the memory requirement needed by the implicit LU algorithm is less than that required by the classical Cholesky factorization.

With regard to the q-banded matrices we show that only $O(q)$ vectors are needed to represent the matrix H_i when the implicit LU, LQ or QR algorithms are applied.

The results that are presented in this chapter can be viewed only as preliminary results in the huge field comprising the analysis of the various types of structured, or unstructured, large linear systems that are met in practice. While leaving to future work a fuller analysis of the ABS methods for this problem (including of course their formulation on vector and parallel computers), it is our opinion that the results presented here already indicate the effectiveness and possibly the competitiveness of using ABS algorithms.

11.2   APPLICATION OF THE IMPLICIT *LU* ALGORITHM TO *ND* MATRICES

In this section we first give some definitions and basic properties of the ND matrices and then we analyse the structure of the matrix H_i generated by the implicit *LU* algorithm. We omit the proofs, given by induction in Zhu (1987a), since they are rather lengthy and technical.

Definition 11.1
Let A be an n by n symmetric positive definite matrix of the form

$$A = \begin{bmatrix} A_1 & G & D \\ G^T & B_1 & E \\ D^T & E^T & F \end{bmatrix} . \tag{11.1}$$

Then A is called a 1-block ND matrix if $G = 0$ and both A_1 and B_1 are square submatrices. A_1 and B_1 are called square blocks and C_1 is called a rectangular block, where C_1 is the submatrix given by

$$C_1 = (D^T, E^T, F) . \tag{11.2}$$

Definition 11.2
Let A be an n by n symmetric positive definite matrix of the form (11.1). Then A is called an m-block matrix if $G = 0$ and both A_1 and B_1 are $(m-1)$-block ND matrices. Besides the square and rectangular blocks in A_1 and B_1, C_1 defined by (11.2) is called a rectangular block.

ND matrices are obtained when the ordering ND algorithm of George (1973) is applied to symmetric positive definite matrices S of general sparsity pattern in order to reduce the possible fill-in in the Cholesky factors. When S is obtained from a regular finite element mesh of size N, then the number q_1 of the nonzero elements in the Cholesky factors, which can appear only in positions corresponding to the nonzero square and rectangular blocks in the obtained ND matrix, is given by

$$q_1 = \frac{31}{4} (N^2 \log_2 N) + 0(N^2) , \tag{11.3}$$

while the number q_2 of multiplications needed to compute the factors is given by

$$q_2 = \frac{829}{84} N^3 + 0(N^2 \log_2 N) . \tag{11.4}$$

It can be shown that the order of the leading terms in the above formulas is optimal, with respect to all possible orderings and for sequential operations. Note, however, that the optimality of the coefficients 31/4 and 829/84 is not yet established. See George (1973), George and Liu (1981) or Deng (1987) for further results on ND matrices.

Now let H_i be the matrix obtained at the ith step of the implicit LU algorithm on an ND matrix A. Let A be represented as follows:

$$A = \begin{bmatrix} A_1 & 0 & A_5 \\ 0 & A_2 & A_4 \\ A_5^{\mathrm{T}} & A_4^{\mathrm{T}} & A_3 \end{bmatrix}, \tag{11.5}$$

where A_1, A_2 and A_3 are square matrices of order p, q and r respectively $(p + q + r = n)$. Let for nonnegative integers $s, t, s \leqslant t \leqslant n, a_i^{s,t} \in R^{t-s+1}$ be the vector defined by

$$a_i^{s,t} = (A_{i,s}, \ldots, A_{i,t})^{\mathrm{T}} . \tag{11.6}$$

Then the following results are proved by Zhu (1987a).

Lemma 11.1

For $i = 2, \ldots, p + 1$ the matrix H_i has the following structure:

$$H_i = \begin{bmatrix} 0 & 0 & 0 & 0 \\ D_i & I_{p-i+1} & 0 & 0 \\ 0 & 0 & I_p & 0 \\ C_1 & 0 & 0 & I_r \end{bmatrix}, \tag{11.7}$$

where $D_i, C_i \in R^{p-i+1,i-1}$ are given by

$$D_2 = -\frac{1}{A_{1,1}a_1^{2,p}} , \tag{11.8}$$

$$C_2 = -\frac{1}{A_{1,1}a_1^{p+q+1,n}} , \tag{11.9}$$

and, for $i = 2, \ldots, p$,

$$D_{i+1} = (\overline{D}_i - u_i d_i^{\mathrm{T}}, \ -u_i) , \tag{11.10}$$

$$C_{i+1} = (C_i - v_i d_i^{\mathrm{T}}, \ -v_i) , \tag{11.11}$$

where d_i and $\overline{D}_i$ consist respectively of the first row and the last $p - i$ rows of D_i and

$$u_i = \frac{D_i a_i^{1,i-1} + a_i^{i+1,p}}{\mu_i} \quad , \tag{11.12}$$

$$v_i = \frac{C_i a_i^{1,i-1} + a_i^{i+1,p}}{\mu_i} \quad , \tag{11.13}$$

$$\mu_i = d_i^{\mathrm{T}} a_i^{1,i-1} + A_{i,i} \quad . \tag{11.14}$$

Lemma 11.2

For $i = p + 2, \ldots, p + q + 1$, the matrix H_i has the following structure:

$$H_i = \begin{bmatrix} 0 & 0 & 0 & 0 \\ 0 & 0 & 0 & 0 \\ 0 & D_i & I_{p+q+1-i} & 0 \\ C_i & F_i & 0 & I_r \end{bmatrix} , \tag{11.15}$$

where $D_i \in R^{p+q+1-i,i+p-1}$ and $F_i \in R^{r,i+p-1}$ are given by the recursions

$$D_{p+2} = \frac{-a_{p+1}^{p+2,p+q}}{A_{p+1,p+1}} \quad , \tag{11.16}$$

$$F_{p+2} = \frac{-a_{p+1}^{p+q+1,n}}{A_{p+1,p+1}} \quad , \tag{11.17}$$

and, for $i = p + 2, \ldots, p + q$,

$$D_{i+1} = (\overline{D}_i - u_i d_i^{\mathrm{T}}, \quad -u_i) \quad , \tag{11.18}$$

$$F_{i+1} = (F_i - v_i d_i^{\mathrm{T}}, \quad -v_i) \quad , \tag{11.19}$$

where $\overline{D}_i$ and d_i are defined as in Lemma 11.1 and

$$u_i = \frac{\overline{D}_i a_i^{p+1,i-1} + a_i^{i+1,p+q}}{\mu_i} \quad , \tag{11.20}$$

$$v_i = \frac{F_i a_i^{p+1,i-1} + a_i^{p+q+1,n}}{\mu_i} \quad , \tag{11.21}$$

with

$$\mu_i = d_i^T a_i^{p+1, i-1} + A_{i,i} \ .$$
(11.22)

Moreover, with C_{p+1} the matrix defined in Lemma 11.1 is

$$C_i = C_{p+1} \ .$$
(11.23)

Lemma 11.3
For $i = p + q + 2, \ldots, n + 1$, the matrix H_i has the following structure:

$$H_i = \begin{bmatrix} 0 & 0 \\ G_i & I_{n-i+1} \end{bmatrix} ,$$
(11.24)

where the matrix $G_i \in R^{n-i+1, i-1}$ is given by the recursions

$$G_{p+q+1} = (C_{p+q+1}, F_{p+q+1}) ,$$
(11.25)

where C_{p+q+1}, F_{p+q+1} are the matrices defined in Lemma 11.2 and

$$G_{i+1} = (\overline{G}_i - v_i q_i^T, -v_i) ,$$
(11.26)

where

$$v_i = \frac{(\overline{G}_i a_i^{1, i-1} + a_i^{i+1, n})}{\mu_i}$$
(11.27)

$$\mu_i = g_i^T a_i^{1, i-1} + A_{i,i}$$
(11.28)

and g_i and $\overline{G}_i$ consist of the first and the last $n - i$ rows of G_i respectively.

The following theorem, based upon the given lemmas, characterizes the structure of the matrix H_i.

Theorem 11.1
The matrices H_i generated by the implicit *LU* algorithm applied to an *m*-block ND matrix have the form

$$H_i = \begin{bmatrix} 0 & 0 \\ S_{i-1} & I_{n-i+1} \end{bmatrix} ,$$
(11.29)

where the nonzero elements of $S_{i-1} \in R^{n-i+1, i-1}$ can appear only in the positions corresponding to the nonzero square and rectangular blocks of A.

Remark 11.1
It is clear that the memory requirement for the nonzero elements of S_{i-1} is sensibly less than the memory requirement for the nonzero elements of the Cholesky factors.

11.3   APPLICATION OF THE IMPLICIT *LU*, *QR* AND *LQ* ALGORITHMS TO BANDED-TYPE MATRICES

We start by defining the banded-type matrices that we shall consider in this section.

Definition 11.3
An n by n matrix A is called a q-band matrix if $A_{i,j} = 0$ for $|i - j| > q$ while $A_{i,j} \neq 0$ for at least one pair i, j such that $|i - j| = q$, $1 \leq i, j \leq n$.

Definition 11.4
An n by n matrix A is called an upper q-band matrix if $A_{i,j} = 0$ for $j > i + q$, while $A_{i,i+q} \neq 0$ for at least one i, $1 \leq i \leq n$.

Definition 11.5
An n by n matrix A is called an upper q/p-band matrix if $A_{i,j} = 0$ for $j > i + q$, while $A_{i,i+q} \neq 0$ for at least one i, $1 \leq i \leq n - p$ and $A_{i,p} \neq 0$ for at least one i, $1 \leq i \leq n - q$.

Definition 11.6
A q-band matrix is said to be t-cyclic if there are indexes j_k, $0 \leq k \leq t \geq 1$, $0 = j_1 \leq \ldots \leq j_t = q$, such that $A_{i,i+j}$ and $A_{i,i-j}$ are zero for $0 \leq j \leq q$, unless j equals one of the j_k.

Definition 11.7
A q-band matrix is said to be a regular q/p-cyclic matrix if p is a divisor of q and, for $0 \leq k \leq q/p$, $j_k = kp$. A regular q/q-cyclic matrix is also called a q-tridiagonal matrix. A vector $u \in R^n$ is said to be q-cyclic if $u_j = 0$ unless $j = m + kq$ for some m, k, with $1 \leq m \leq q$, $0 \leq k \leq (n-m)/q$.

We consider first the application of the implicit *LU* algorithm on banded matrices of the defined type. We assume that A is strongly nonsingular, to avoid pivoting issues that increase the band width. Given a vector $u \in R^n$, we indicate by $u^{s,q} \in R^{q-s+1}$ the vector comprising the elements of u from the sth to the qth. Also we recall, from Theorem 6.3, that the matrix H_{i+1} has the following form, with $S_i \in R^{n-i,i}$:

$$H_{i+1} = \begin{bmatrix} 0 & 0 \\ S_i & I_{n-i} \end{bmatrix}. \tag{11.30}$$

The following theorem, due to Abaffy and Dixon (1987) and generalizing previous results of Abaffy (1986) and independent results of Phua (1986, 1988), covers the case of q-band and upper q-band matrices.

Theorem 11.2
Let A be an upper q-band matrix. Then the nonzero elements of S_i in (11.30) can only lie in the first q rows of S_i.

Proof
For $i = 1$ we have from (6.12) that $S_1 = a_1^{2,n}$ and the result follows since $a_1^{2+q,n} = 0$. Proceeding by induction, we can assume the following structure for H_i:

$$H_i = \begin{bmatrix} 0 & 0 & 0 \\ C_{i-1} & I_q & 0 \\ 0 & 0 & I_{n-q-i+1} \end{bmatrix}, \tag{11.31}$$

where C_{i-1} is the matrix comprising the first q rows of S_{i-1}. From (11.31) it follows that $(H_i a_i)^{i+q+1,n} = 0$ while $p_i^{i+1,n} = 0$ from (6.11). Thus in S_i the last $n - i - q$ rows are zero and the theorem follows. Q.E.D.

Corollary 11.1
If the kth row of C_{i-1} is zero, $2 \leqslant k \leqslant q$, and $A_{i,i+q}$ is also zero then the $(k-1)$th row of C_i is zero.

We omit the proof of the following theorem, which is similar to that of Theorem 11.2.

Theorem 11.3
Let A be an upper q/p-band matrix. Then the nonzero elements of S_i lie only in the first q and in the last p rows of S_i.

Remark 11.2
Theorem 11.3 can be easily generalized to the case of q-band matrices with extra nonzero elements appearing above the diagonal in p columns of indexes $j_1, \ldots, j_p$. Then the nonzero elements of S_i lie in the first q rows and in those rows of S_i which consist of the first i elements of the j_1th, $\ldots$, j_pth rows of H_{i+1}.

Remark 11.3
The maximum storage requirement for S_i in the case of an upper q-band matrix is $q(n-q)$; therefore less than half the storage required by the classical LU factorization (i.e. $2qn + n - q^2 - q$). For a q-band matrix the number of multiplications and divisions is no more than $n(nq + q^2 + 2)/2 - q(5q^2 + 3q + 4)/6 + n - 1$, which for $n \gg q$ is dominated by $n^2q/2$. This bound is one order greater, for $n \gg q$, than the corresponding number for the classical LU factorization, which is (Pissanetzky 1984) $(q + 1)qn - 2q^3/3 - q^2 - q/3$. For an upper q-band matrix, the number of multiplications is $n^2q - 2nq^2 + 4q^2/3 + O(n^2, nq, q^2)$.

By induction, one can easily prove the following.

Theorem 11.4
Let A be a regular q/p cyclic matrix. Then the nonzero elements of S_i are only those elements of the form $(S_i)_{k,i+k-jq}$, with $1 \leqslant k \leqslant q$ and $1 \leqslant j \leqslant (i+q)/q$. The vector comprising the first $i-1$ components of p_i is q-cyclic.

Continuing this type of analysis, we can study the effect of any nonzero elements of the upper triangular part of A, as done in the following theorem, whose proof can be given along similar lines as above.

Theorem 11.5
When the implicit LU method is used, if the element $A_{k,j}, j > k$, is nonzero, then the row of C_i corresponding to the jth row of H_i is nonzero if $j > i > k$.

Remark 11.4
Using Theorem 11.5, one can predict where the nonzero elements of C_i are located. The simple structure of H_i in many frequently occurring cases is a consequence of Theorem 11.5.

We consider now the application of the implicit QR algorithm on a q-band nonsingular matrix A. Recall that the implicit QR algorithm is well defined without need for pivoting and that it can be viewed as the implicit LU algorithm applied to the problem with coefficient matrix $P^T A^T A$. It generates therefore matrices H_i of the form (11.30).

Theorem 11.6
Let A be a q-band matrix. Then the nonzero elements of S_i in the matrix H_{i+1} generated by the implicit QR algorithm lie only in the first $2q$ rows of S_i.

Proof
Since the implicit QR algorithm is just the implicit LU algorithm applied to the problem with coefficient matrix $P^T A^T A$, it is enough to prove that $P^T A^T A$ is an upper $2q$-band matrix, and Theorem 11.6 follows then from Theorem 11.2. Since $P^T A^T$ is the product of a lower triangular by a q-band matrix, it is an upper q-band matrix. Since the product of an upper q-band matrix by an s-band matrix is an upper $(q + s)$-band matrix, the theorem is proved Q.E.D.

When A is a regular cyclic q/p-band matrix, similar results to those in Theorems 11.4 and 11.5 can be established for the implicit QR algorithm. See Abaffy and Dixon (1987) for details.

We consider finally the application of the Huang algorithm to banded type matrices. We need the following.

Definition 11.8
An n by n matrix A is called a q_1/q_2-band matrix if $A_{i,j} = 0$ for $i > j + q_1$ while $A_{j,j+q_1} \neq 0$ for at least one j, $1 \leqslant j \leqslant n$, and $A_{i,j} = 0$ for $j > i + q_2$, while $A_{i,i+q_2} \neq 0$ for at least one i, $1 \leqslant i \leqslant n$.

Theorem 11.7

Let A be a q_1/q_2-band matrix and consider the matrices H_i generated by the Huang algorithm. Then for the computation of the search vectors it is sufficient to store the first $i + q_2 - 1$ elements of the columns from the $(i - q_1)$th to the $(i + q_2 - 1)$th.

Proof

We proceed by induction, assuming that H_i has the form

$$H_i = \begin{bmatrix} B_i & 0 \\ 0 & I_{n-i-q_2+1} \end{bmatrix}, \tag{11.32}$$

where $B_i \in R^{i+q_2-1, i+q_2-1}$ and observing that (11.32) is true for $i = 1$. Since a_i has nonzero elements only in the positions from the $(i + q_1)$th to the $(i + q_2)$th, the vector $H_i a_i$ can have nonzero elements only in the first $i + q_2$ positions. Thus the correction to H_i affects only the principal submatrix of dimension $i + q_2$ and H_{i+1} has the required structure. The theorem follows observing that the first $i - q_1$ components of a_{i+1} and of all successive rows are zero, so that it is not necessary to store the first $i - q_1$ columns of H_{i+1} for the computation of p_{i+1} Q.E.D.

For additional results, similar to the results in the *LU* case, on the Huang algorithm applied to banded-type matrices (including q-tridiagonal matrices, q/p-band matrices and other frequently occurring types) see Abaffy and Dixon (1987).

11.4   BIBLIOGRAPHICAL REMARKS

The first results on the application of the implicit *LU* and *QR* algorithm on banded-type matrices are due to Abaffy (1986) and independently to Phua (1986, 1988) who considered only the *LU* algorithm. The results given here for banded-type matrices are due to Abaffy and Dixon (1987). The application of the implicit *LU* algorithm to ND matrices is due to Zhu (1987a). Preliminary results on the application of ABS algorithms to problems with arbitrary sparsity pattern are given by Fragnelli and Resta (1985) and Zeng (1988).

12

Error analysis in the scaled ABS algorithm

12.1 INTRODUCTION

In this chapter we analyse the scaled ABS algorithm from the point of view of sensitivity to errors and propagation of round-off errors. The sensitivity to errors in the estimate of the solution and in the residual is investigated through the approach developed by Broyden (1974) for general iterative processes and later applied by Broyden (1985) to the unscaled ABS algorithm. Here we derive a bound to the perturbation induced by a single error for the scaled ABS class and we establish a necessary and sufficient condition (the Broyden condition) for minimizing the derived bound. This condition is satisfied, *inter alia*, by the Huang algorithm as observed in chapter 8, when formulated with the parameter choices $z_i = e_i$, $v_i = A^{-T}p_i$, w_i a multiple of z_i and $H_1 = I$. We prove here the additional result that an error in x_i is propagated in x_{i+1} with no growth, in contrast with the general case, where the growth can be at least linear in the condition number of A.

The propagation of the round-off errors is investigated by a forward error analysis approach. The resulting formulas are, as usually, rather complex and unduly pessimistic. Indications are, however, provided in favour of algorithms of the scaled Huang and the modified scaled Huang subclass.

12.2 SENSITIVITY ANALYSIS VIA THE BROYDEN APPROACH

Broyden (1974) used the following approach to analyse the sensitivity of an iterative process to errors (or perturbations) introduced in a certain step. Let the relation between two iterates x_i, x_{i+1} be expressed by the linear operator

$$x_{i+1} = \Phi_i(x_i) \ . \tag{12.1}$$

Suppose that an error q_i occurs at the ith step, so that x_i is substituted by $x_i' = x_i + q_i$, while no more errors occur in the following iterations. Let, for a certain index $p \geq i$,

x_{p+1} be the value computed in the absence of errors and x'_{p+1} the value computed when the error q_i is made. Then we have

$$x'_{p+1} - x_{p+1} = \Omega^{p-i+1} q_i \;, \tag{12.2}$$

where Ω^{p-i+1} is the linear operator defined by

$$\Omega^{p-i+1}(x) = \Phi_p(\Phi_{p-1}(\ldots(\Phi_i(x))\ldots)) \;. \tag{12.3}$$

The operator Ω^{p-i+1} may contain parameters depending on both the problem and the particular considered algorithm. The Broyden approach consists in selecting the parameters related to the algorithm (but possibly also some parameters related to the problem and available to changes, such as unit lengths) so that Ω^{p-i+1} is as small as possible in some sense.

Assuming now that x is a vector in R^n, so that Φ_i is a matrix, from (12.2) we obtain

$$\|x'_{p+1} - x_{p+1}\| \leq \|\Omega^{p-i+1}\| \, \|q_i\| \;. \tag{12.4}$$

Hence we are led to the following.

Definition 12.1
An iteration of the form (12.1) is said to be p-optimal according to the Broyden criterion with respect to a certain norm $\|.\|$ if $\|\Omega^{p-i+1}\|$ is minimized with respect to any available parameter.

It must be stressed that the Broyden optimality criterion may not be strong enough to characterize the algorithms with actual better average behaviour in a set of problems. One cannot indeed underplay the limitations induced by dealing with a bound and by disregarding the effects of multiple and correlated errors.

We shall now apply the Broyden criterion to the iterates x_i and r_i generated by the scaled ABS algorithm. First we consider x_i. For the scaled ABS algorithm, equation (12.1) takes the form

$$x_{i+1} = (I - S_i)x_i + d_i \;, \tag{12.5}$$

where S_i is the matrix defined in (8.75) and d_i is given by

$$d_i = \frac{v_i^T b}{p_i^T A^T v_i} \, p_i \;. \tag{12.6}$$

Since the scaled ABS algorithm has finite termination, it is natural to apply Broyden criterion with the index $p = n$. Then repeated application of (12.5) gives

$$x_{n+1} = \prod_{j=0}^{n-1}(I - S_{n-j})x_i + \sum_{j=1}^{n}\left(\prod_{k=0}^{n-i+1}(I - S_{n-k})\right)d_j \ . \tag{12.7}$$

Thus (12.4) takes the form

$$\|x'_{n+1} - x_{n+1}\| \leq \|Q^{n,i}\| \, \|x'_i - x_i\| \ , \tag{12.8}$$

where $Q^{n,i}$ is defined by

$$Q^{n,i} = \prod_{j=0}^{n-i}(I - S_{n,j}) \ . \tag{12.9}$$

By Definition 12.1 a method in the scaled ABS class is n-optimal according to the Broyden criterion if $\|Q^{n,i}\|$ is minimal for all i. In order to derive the minimality conditions we need the following results.

Theorem 12.1
The matrix $Q^{n,i}$ given by (12.9) is a projection matrix whose range and null space are defined by

$$\text{Range}(Q^{n,i}) = {}^{\perp}\text{Span}(A^Tv_i, \ldots, A^Tv_n) \ , \tag{12.10}$$

$$\text{Null}(Q^{n,i}) = {}^{\perp}\text{Span}(A^Tv_1, \ldots, A^Tv_{i-1}) \ . \tag{12.11}$$

Proof
Using equation (8.76) in Theorem (8.21) we find that $S_jQ_{n,i} = 0$ for $i \leq j \leq n$ and that $(I - S_j)Q^{n,i} = Q^{n,i}$ for $i \leq j \leq n$, implying that $Q^{n,i}$ is idempotent and hence a projector. If C, D and CD are projectors, it is known that $\text{Range}(CD) = \text{Range}(C) \cap \text{Range}(D)$ and that $\text{Null}(CD) = \text{Null}(C) \cup \text{Null}(D)$. Observing that $\text{Range}(I - S_i) = \text{Null}(S_i)$ and that $\text{Null}(I - S_i) = \text{Range}(S_i)$, we obtain

$$\text{Range}(Q^{n,i}) = \bigcap_{j=i}^{n}\text{Null}(S_j) = \bigcap_{j=i}^{n}{}^{\perp}\text{Range}(A^Tv_j) = {}^{\perp}\text{Span}(A^Tv_i, \ldots, A^Tv_n)$$

and that

$$\text{Null}(Q^{n,i}) = \bigcup_{j=i}^{n}\text{Range}(S_j) = {}^{\perp}\text{Span}(A^Tv_1, \ldots, A^Tv_{i-1}) \ .$$

Q.E.D.

Lemma 12.1
Let A be an idempotent matrix with rank $q \geq 1$. Then the following inequality is true:

$$\|A\|_F \geq q^{1/2} \tag{12.12}$$

and the equality sign holds if and only if A is symmetric.

Proof
It is known that the Frobenius norm is unaffected by orthogonal transformations and that an orthogonal matrix U exists such that $U^T A U = B$, where B has the form

$$B = \begin{bmatrix} I_q & C \\ 0 & 0 \end{bmatrix} \tag{12.13}$$

and $C \in R^{q, n-q}$ is some matrix. Thus we have $\|A\|_F = \|B\|_F \geq q^{1/2}$, with the equality sign if and only if $C = 0$, i.e. if and only if B is symmetric. Since B is symmetric if and only if A is so, the lemma follows. Q.E.D.

Lemma 12.2
If A is idempotent and nonzero, then $\|A\|_2 \geq 1$ and the equality sign holds if and only if A is symmetric.

Proof
It is known that the spectral norm is invariant under orthogonal transformations. Thus, defining B as in Lemma 12.1, we have $\|A\|_2 = \|B\|_2 = [\mu(B^T B)]^{1/2}$. Now $B^T B$ has the form

$$B^T B = \begin{bmatrix} I_q & C \\ C^T & C^T C \end{bmatrix}. \tag{12.14}$$

Defining the matrix L by

$$L = \begin{bmatrix} I_q & 0 \\ C^T & I_{n-q} \end{bmatrix} , \tag{12.15}$$

we easily verify that L^{-1} is obtained by changing C^T to $-C^T$ in (12.15) and that the similarity transformation $L^{-1} B^T B L$ yields

$$L^{-1} B^T B L = \begin{bmatrix} I_q + C^T C & C \\ 0 & 0 \end{bmatrix} . \tag{12.16}$$

Thus $B^T B$ has $n - q$ zero eigenvalues and q eigenvalues given by the equation

$\det[CC^T - (\sigma - 1)I_q] = 0$. If $C \neq 0$, then CC^T is positive semidefinite; hence its eigenvalues are nonnegative and at least one of them is positive. Thus there is an eigenvalue of B^TB which is greater than one, implying that $\mu(B^TB) > 1$. For symmetric B (or A) it is obvious that $\|B\|_2 = 1$. Q.E.D.

We can now establish the main result.

Theorem 12.2
An algorithm of the scaled ABS class is n-optimal according to the Broyden criterion with respect to the iterate x_i and to the spectral or the Frobenius norms if and only if the following condition holds, with D a diagonal matrix:

$$V^TAA^TV = D .\tag{12.17}$$

Proof
From Lemmas 12.1 and 12.2 it follows that the required necessary and sufficient conditions are that $Q^{n,i}$ be symmetric. Since a projector P is symmetric if only only if $\text{Range}(P) = {}^\perp\text{Null}(P)$, it follows from Theorem 12.1 that $Q^{n,i}$ is symmetric if and only if $\text{Span}(A^Tv_1, \ldots, A^Tv_{i-1}) = {}^\perp\text{Span}(A^Tv_i, \ldots, A^Tv_n)$, which is equivalent to $(A^Tv_k)^T(A^Tv_j) = 0$ for $k \neq j$, or, in matrix form, to (12.17). Q.E.D.

Remark 12.1
As already observed in Chapter 8, condition (12.17) was originally derived by Broyden (1985) and is satisfied by all algorithms in subclass S4 (where $v_i = A^{-T}p_i$ and the p_i are orthogonal).

Remark 12.2
Condition (12.17) does not apply for the Huang algorithm in its standard formulation ($v_i = e_i$), but the Huang algorithm is however equivalent to a method in subclass S4 (see Theorem 8.25). More precisely, since the algorithms of subclass S4 generate the same sets of iterates x_i as the algorithms of subclass S1 (see Theorem 8.26), it follows that there are algorithms in the scaled ABS class which do not satisfy the Broyden optimality criterion but which behave identically as algorithms that satisfy that criterion.

We now derive a bound to $\|Q^{n,i}\|$ in terms of condition numbers. It is convenient to use the Householder notation (see section 1.3).

Theorem 12.3
For both the Frobenius and the spectral norms of the matrix $Q^{n,i}$ defined in (12.9) the following inequality holds:

$$\|Q^{n,i}\| \leq \text{Cond}(V^TA) .\tag{12.18}$$

Proof

Since $Q^{n,i}$ is a projector onto $\mathrm{Range}(A^{\mathrm{T}}V)^{n-i|}$ along $\mathrm{Range}(A^{\mathrm{T}}V)^{|i-1}$, it can be represented in the form

$$Q^{n,i} = (A^{-1}V^{-\mathrm{T}})^{|i-1}(V^{\mathrm{T}}A)^{\overline{i-1}} . \tag{12.19}$$

Since, for $i \leqslant k \leqslant n$, $\|B^{|k|}\|$ and $\|B^{\bar{k}}\|$ are bounded by $\|B\|$ in both the Frobenius and the spectral norms, it follows from (12.19) that $\|Q^{n,i}\| \leqslant \|A^{-1}V^{-\mathrm{T}}\| \|V^{\mathrm{T}}A\|$. Q.E.D.

Remark 12.3

If the matrix V contains parameters u (such as scaling factors in p_i, v_i, A) which do not affect S_i, then (12.18) can be substituted by the stronger inequality

$$\|Q^{n,i}\| \leqslant \inf_u \mathrm{Cond}[V(u)^{\mathrm{T}}A] . \tag{12.20}$$

From (12.18) we can bound the perturbation in x_{n+1} by

$$\|x'_{n+1} - x_{n+1}\| \leqslant \mathrm{Cond}(V^{\mathrm{T}}A)\|q_i\| \tag{12.21}$$

or by

$$\|x'_{n+1} - x_{n+1}\| \leqslant \mathrm{Cond}(V)\,\mathrm{Cond}(A)\|q_i\| . \tag{12.22}$$

Broyden (1985) investigated by a different technique the unscaled ABS class obtaining the inequality

$$\|x'_{n+1} - x_{n+1}\| \leqslant \mathrm{Cond}(A)\|q_i\| . \tag{12.23}$$

Since the scaled ABS algorithm is equivalent to the unscaled ABS algorithm applied to the modified problem $V^{\mathrm{T}}Ax = V^{\mathrm{T}}b$, (12.21) follows directly from Broyden's result.

From (12.22) it follows that (12.23) is valid not only when $V = I$, but also when V is orthogonal (or has orthogonal columns, by using (12.20)). Thus it applies for subclass S3. When (12.23) holds, the error propagation due to the algorithm is comparable with the perturbation in the solution induced by a change in the matrix or in the right-hand side.

If we consider subclass S4, then no growth of error is produced, as shown by the following theorem.

Theorem 12.4
For all algorithms in subclass S4 the following inequality is true in the Broyden model of error propagation:

$$\|x'_{n+1} - x_{n+1}\|_2 \leqslant \|q_i\|_2 \ . \tag{12.24}$$

Proof
In subclass S4 we have $V = A^{-T}P = A^{-T}Q^T D$, where Q is orthogonal and D diagonal (see Theorem 8.23). Hence $\text{Cond}(V^T A) = \text{Cond}(DQ)$ and (12.24) follows from (12.20) since $\inf_D[\text{Cond}(DQ)] = \text{Cond}(Q) = 1$. Q.E.D.

While Theorem 12.4 gives a strong theoretical indication that the algorithms in subclass S4, and the equivalent algorithms in subclass S1, should have good numerical stability also in practice, we again stress that there are limitations in the Broyden model. The computational experience presented in Chapter 10 indicates, for instance, that different implementations of the Huang algorithm have different numerical behaviour and in particular that the modified Huang algorithm is more stable than the standard Huang algorithm, a fact which cannot be detected via this approach.

We consider now the sensitivity in the residual induced by a single error at the ith iteration. Since $r_i - r'_i = A(x_i - x'_i)$, we obtain, by repeated application of (12.5),

$$r_{n+1} - r'_{n+1} = AQ^{n,i}A^{-1}(r_i - r'_i) \ , \tag{12.25}$$

implying the bound

$$\|r_{n+1} - r'_{n+1}\| \leqslant \|AQ^{n,i}A^{-1}\| \, \|r_i - r'_i\| \ . \tag{12.26}$$

The condition under which $\|AQ^{n,i}A^{-1}\|$ is minimal is given in the following theorem.

Theorem 12.5
An algorithm of the scaled ABS class is n-optimal according to the Broyden criterion with respect to the iterate r_i and both the Frobenius and the spectral norms if and only if the following condition holds, with D a diagonal matrix

$$V^T V = D \ . \tag{12.27}$$

Proof
If A_1, A_2, A_3, A_4 are square matrices, they satisfy the identity

$$(A_1 A_2)^{|k}(A_3 A_4)^{\bar{k}} = A_1(A_2^{|k}A_3^{\bar{k}})A_4 \ . \tag{12.28}$$

Using (12.28) and (12.19), we obtain

$$AQ^{n,i}A^{-1} = (V^{-T})^{|i-1|}(V^T)^{\overline{i-1}} , \tag{12.29}$$

showing that $AQ^{n,i}A^{-1}$ is a projector onto $\text{Range}[(V^T)^{|i-1|}]$ along $\text{Range}[(V^{-T})^{n-i+1|}]$. Thus $\|AQ^{n,i}A^{-1}\|$ is minimal if and only if $\text{Range}[(V^{-T})^{|i-1|}] = {}^\perp\text{Range}[(V^{-T})^{n-i+1|}]$ for all i. This condition is satisfied if and only if $(V^{-T})^T(V^{-T}) = D^{-1}$ for a diagonal matrix D, implying (12.27). Q.E.D.

Remark 12.4
Condition (12.27) is satisfied by all algorithms in the unscaled ABS class and in subclass S3. The Huang algorithm therefore satisfies the Broyden optimality criterion with respect to errors in both x_i and r_i.

We note that preliminary results on the sensitivity to errors in H_i and on the use of a model which allows errors in each iterations are given by Galantai (1987).

12.3   FORWARD ERROR ANALYSIS IN FLOATING-POINT ARITHMETIC
In this section we present bounds for the error in p_i and x_i when floating-point arithmetic is used. The bounds are, as usually, given by rather complex formulas, which may additionally be unduly pessimistic. They might be improved and simplified for particular algorithms, but we do not offer here a particular analysis of specific algorithms. We also give only the final formulas, omitting their derivation, which is rather cumbersome and can be found in the paper by Abaffy *et al.* (1989).
In the following, Euclidean norms are used for vectors and Frobenius norms for matrices. By $O'(a)$ we indicate a quantity (scalar, vector or matrix) whose norm is bounded by $\|a\|$. By $fl(a)$ we indicate the computed value by floating-point arithmetic of a quantity whose exact value is a. In the derivation of the bounds, systematic use is made of the following relation (Forsythe and Moler 1967):

$$fl(x^Ty) = x^Ty + O'(\phi(n)\varepsilon\|x\|\ \|y\|) \tag{12.30}$$

where ε is the machine zero $(fl(1 + \varepsilon) = 1)$ and $\phi(n) \le 1.01n$.
First we deal with the error propagation in the search vector. We write the update of H_i in the form

$$H_{i+1} = H_i - \frac{H_iA^Tv_iu_i^TH_i}{u_i^TH_iA^Tv_i} , \tag{12.31}$$

and we assume that p_{i+1} is computed through the following sequence of computations, operations in the inner parenthesis being performed first: A^Tv_i, $H_i(A^Tv_i)$, $H_i^Tu_i$, $(H_i^Tu_i)^T(A^Tv_i)$, $s_i = H_i^Tu_i/(H_i^Tu_i)^T(A^Tv_i)$, $D_i = (H_iA^Tv_i)s_i^T$, $H_{i+1} = H_i - D_i$, $p_{i+1} = H_{i+1}^Tz_{i+1}$.
We also make the following assumptions for all i.

(a) The vectors z_i, v_i, u_i have unit norms and the angle between $H_i^T u_i$ and $A^T v_i$ is less than $\tau < \pi/2$.

(b) If δ_i is given by

$$\delta_i = \frac{\|H_i\|}{\|H_i^T u_i\|} ,$$

(12.32)

then there is a scalar β such that

$$\varepsilon \phi(n) < \frac{(\beta - 1) \cos \tau}{3 \beta \delta_i \, \mathrm{Cond}\,(A)}$$

(12.33)

and

$$\varepsilon \phi(n) \ll \frac{1}{4 \beta \delta_i \, \mathrm{Cond}(A)} .$$

(12.34)

Condition (a) implies no loss of generality, since the sequences x_i and H_i are not affected (in exact arithmetic) by the norms of z_i, v_i, u_i and the iteration is finite. Condition (b) is violated if the condition number is sufficiently large (in such a case, higher-order terms in ε, which can be ignored under condition (b), must be taken into account).

Under conditions (a) and (b) the following bound is obtained for the accumulated error in p_{i+1}:

$$\|fl(p_{i+1}) - p_{i+1}\| \leq \varepsilon C_{i+1} .$$

(12.35)

The growth factor C_{i+1} is given by

$$C_{i+1} = \phi(n)\|H_{i+1}\| + \sum_{j=1}^{i} \theta_j [1 + \mu(1 + \delta_j + \|H_j\|)]^{i-j} ,$$

(12.36)

where

$$\mu = \frac{1}{\cos \tau}$$

(12.37)

and

$$\theta_j = \|H_{j+1}\| + 2\phi(n)\|H_j\| \, \|A\| \, \|s_j\|$$

$$\frac{+ \; \|H_j A^T v_j\|(\|s_j\| + \mu\{1 + \phi(n)\delta_j[1 + 3\beta\mu\,\mathrm{Cond}(A)]\})}{\|A^T v_j\|} \; . \tag{12.38}$$

To study the error propagation in x_i, we use (12.34), which can be written in the form

$$fl(p_i) = p_i + O'(\varepsilon C_i) \; . \tag{12.39}$$

In addition to conditions (a) and (b) we assume the validity of the following conditions, the first of which allows to neglect terms which are of order ε^2:

$$C_i \ll \min\left(\frac{1/\varepsilon - \phi(n)}{1 + \varepsilon\phi(n)} \; , \; \frac{1/\varepsilon - 1}{1 + \varepsilon}\right) \tag{12.40}$$

and, with δ some scalar greater than one,

$$C_i < \frac{(\delta - 1)\cos\tau}{\delta\varepsilon} - \phi(n)[1 + \mathrm{Cond}(A)] \; . \tag{12.41}$$

Then we can write for the propagated error in x_{i+1}

$$\|fl(x_{i+1}) - x_{i+1}\| \leqslant \varepsilon D_i \; , \tag{12.42}$$

where, the growth function D_{i+1} satisfies

$$\begin{aligned} D_{i+1} \leqslant [1 + \mu\mathrm{Cond(A)}]D_i + |\alpha^i|\{(1 + \delta\mu)C_i + 2 \\ + \delta\mu\phi(n)[1 + \mathrm{Cond}(A)]\} + \|x_{i+1}\| + \|A^{-1}\|\mu\|r_i\|[1 + \phi(n)] \\ + \phi(n)\mu\,\mathrm{Cond}(A)\|x_i\| \; . \end{aligned} \tag{12.43}$$

From (12.43) an expression for the accumulated error in x_{i+1} can be derived (Abaffy *et al.* 1989).

From an inspection of equations (12.36) and (12.43) we see that terms which can make the growth function larger and which can be controlled by parameter choices are μ, $\|H_i\|$, $\|x_i\|$. If we consider the algorithms in the scaled Huang subclass, then, by Lemma 12.2, $\|H_i\|_2$ equals one and, by Theorem 9.4, $\|x_i\|_2$ is bounded by $\|x^+\|_2$, so that these two factors cannot be arbitrarily large. If moreover we define the scalar product in the denominator of the definition of s_i not as $(H_i^T u_i)^T (A^T v_i)$, but as $u_i^T(H_i A^T v_i)$, then condition (a) becomes that the angle between u_i and $H_i A^T v_i$ be less than $\tau < \pi/2$. If we take $u_i = H_i A^T v_i$, then $\tau = 0$, or μ has the minimum value one.

This choice for u_i corresponds to the scaled modified Huang update, whose numerical superiority is again suggested by the floating-point error analysis.

12.4 BIBLIOGRAPHICAL REMARKS

The results presented in section 12.2 are due to Galantai (1987) except Theorem 12.4 which is due to Spedicato. The results in section 12.3 are mainly due to Abaffy and Galantai and are available in the paper by Abaffy *et al.* (1989). Further results on error propagation for ABS methods are available in a paper by Abaffy (1987a).

13

The ABS algorithm for nonlinear systems

13.1 INTRODUCTION

In this chapter we consider a class of algorithms, based upon the scaled block ABS algorithm introduced in Chapter 7, for finding a local zero x^+ of the nonlinear system

$$f(x) = 0 , \tag{13.1}$$

where $x \in R^n$ and $f(x) = (f_1(x), \ldots, f_n(x))^T$ is a continuous mapping from R^n to R^n. The class that we give is very general. It contains as special cases the Newton method, the Stewart (1973) nonlinear generalized conjugate direction methods, the methods of Brown (1969) and Brent (1973) and their generalization by Gay (1975a) and Cosnard (1975).

As established by Theorem 13.1 the methods in the ABS class are locally Q-superlinearly convergent under the standard assumptions on $f(x)$ and some conditions on the free parameters. The algorithms corresponding to the orthogonally scaled and to the optimally stable subclasses defined in Chapter 8, and their block generalization, are locally Q-superlinearly convergent. These subclasses include the Huang algorithm and the modified Huang algorithm (which are equivalent to the Brent algorithm), the implicit QR algorithm and the minimum AA^T-iterations and minimum A^TA-iterations algorithms of the conjugate gradient type.

Some numerical experiments are presented, which show that a straightforward implementation of the modified Huang algorithm has a performance virtually identical with the sophisticated implementation due to Moré and Cosnard (1979) of the Brent algorithm.

In the following we indicate by $A(x)$ the Jacobian of $f(x)$, i.e. the matrix whose elements are defined by

$$A_{i,j}(x) = \frac{\delta f_i(x)}{\delta x_j} . \tag{13.2}$$

By $a_i^{\mathrm{T}}(x)$, $i=1, \ldots, n$, we indicate the ith row of the Jacobian. The convergence proof requires that $A(x^+)$ is nonsingular and Lipschitz continuous, implying non-singularity of $A(x)$ in a neighbourhood of x^+. For simplicity of formulation we shall additionally assume the nonsingularity of the matrix $(A^{\mathrm{T}}(u_1)V_1, \ldots, A^{\mathrm{T}}(u_k)V_k)$, where $1 \leqslant k \leqslant r \leqslant n$, $V=(V_1, \ldots, V_k)$ is nonsingular and $u_1, \ldots, u_k$ are vectors defined in the inner cycle of the nonlinear ABS algorithm.

13.2 FORMULATION OF THE SCALED BLOCK NONLINEAR ABS ALGORITHM

The scaled block nonlinear ABS algorithm can be considered as a generalization of the Newton method aiming at using more recent information in the same way as the Gauss–Seidel process does compared with the Jacobi process. The Newton method in its basic iteration has the form

$$x_{i+1} = x_i - d_i \; , \tag{13.3}$$

where $d_i \in R^n$ solves the linear system

$$A(x_i)d_i = f(x_i) \; . \tag{13.4}$$

The idea motivating the nonlinear ABS method is the following. Let x_i be a current approximation of x^+. Consider instead of the system (13.4) a number $r(i)$, $1 \leqslant r(i) \leqslant \mu$, of systems obtained by partitioning the components of the mapping f and the rows of the Jacobian into $r(i)$ blocks and evaluating them at different dynamically defined points, say

$$A_k(u_k)d_k = f_k'(y_k) \; , \qquad k=1, \ldots, r(i) \; , \tag{13.5}$$

where A_k consists of some rows of A and f_k' of the corresponding mapping components, and u_k and y_k are defined as follows: for $k=1$ the selected rows of the Jacobian and components of the mapping are both evaluated in $u_1 = y_1 = x_i$. A solution d_1 of the system is computed by taking a step of the block ABS algorithm starting with the zero vector. Consider now the vector $y_2 = x_i - d_1$. In view of the fact that d_1 satisfies the first of the Newton equations, y_2 may be considered in a sense as a better approximation of the solution than x_i. Define now the second system by evaluating f_2' at y_2 and A_2 at an arbitrary convex combination u_2 of y_1 and y_2. Determine a solution d_2 of both the second system and the first system by taking a second step of the block ABS algorithm. Set $y_3 = x_i - d_2$ and define u_3 as an arbitrary convex combination of y_1, y_2 and y_3. By proceeding thus, a vector $d_{r(i)}$ is finally obtained which solves the $r(i)$ dynamically defined systems. Then x_{i+1} is defined as $x_{i+1} = x_i - d_i = y_{r(i)+1}$.

By such a procedure, the original Newton iteration has been transformed into a cycle of $r(i)$ minor iterations. Each minor iteration provides a better approximation of the solution, which is exploited in the evaluation of the next block into which the

Newton system is decomposed. Note that, if $r(i) = 1$, then the Newton method is obtained.

The vectors x_i are called the major iterates. The iterates defined inside the ith cycle are called the ith cycle minor iterates. Unless confusion may arise, we shall omit for simplicity of notation the index i which should be attached to the minor iterates.

We formalize now the procedure described relating to the ith cycle. It is assumed that an initial estimate x_1, sufficiently close to the solution, is given.

ALGORITHM 15: The Scaled Block Nonlinear ABS Cycle

(A15) Set $y_1 = x_i$, $H_1 = I$. Let $r(i)$ be an integer in the range $[1, n]$. Let $m_1, \ldots, m_{r(i)}$ be integers greater or equal to one and such that $m_1 + \ldots + m_{r(i)} = n$. Set $k = 1$.

(B15) Define the vector $u_k \in R^n$ by

$$u_k = \sum_{j=1}^{k} \tau_{k,j} y_j ,$$ (13.6)

where the scalars $\tau_{k,j}$ are arbitrary nonnegative numbers whose sum equals one.

(C15) Define the n by m_k matrix P_k by

$$P_k = H_k^T Z_k ,$$ (13.7)

where Z_k is an arbitrary n by m_k matrix subject to

$$\det[P_k^T A^T(u_k) V_k] \neq 0$$ (13.8)

and V_k is an n by m_k matrix arbitrary save that $(V_1, \ldots, V_k)$ is full rank.

(D15) Define the vector $d_k \in R^n$ by

$$d_k = [V_k^T A(u_k) P_k]^{-1} V_k^T f(y_k) .$$ (13.9)

(E15) Update y_k by

$$y_{k+1} = y_k - P_k d_k .$$ (13.10)

If $k = r(i)$, set $x_{i+1} = y_{r(i)+1}$; the cycle is completed.

(F15) Update H_k by

$$H_{k+1} = H_k - H_k A(u_k)^T V_k W_k^T H_k ,$$ (13.11)

where W_k is an n by m_k matrix, arbitrary save that

$$W_k^T H_k A(u_k)^T V_k = I_{m_k} . \tag{13.12}$$

Increment the index k by one and go to (B15).

Remark 13.1

If $f(x)$ is a linear mapping, $f(x) = Ax - b$, then the above cycle is equivalent to ALGORITHM 6 and $y_{r(i)+1}$ is the solution for any starting y_1. If only the first j components of $f(x)$ are linear, they are still solved exactly by $y_{r(i)+1}$. See Schmidt and Hoyer (1978) and Hoyer (1981) for a discussion of how to exploit linearities in a mapping in the framework of the Brown and the Brent methods.

13.3 LOCAL CONVERGENCE OF THE SCALED BLOCK NONLINEAR ABS ALGORITHM

In this section we give a local convergence theorem for the scaled block nonlinear ABS algorithm defined in the previous section (ALGORITHM 15). Note that we use Euclidean norms for vectors and Frobenius norms for matrices. By $U(x^+, \delta_0)$ we indicate the open Euclidean ball of centre x^+ and radius δ_0 (the set of points x such that $\|x - x^+\| < \delta_0$).

The theorem is established under the following conditions on the mapping and its Jacobian.

(a) The mapping $f(x)$ has an isolated zero x^+
(b) The Jacobian $A(x^+)$ is nonsingular.
(c) There exist two positive scalars δ_0, τ_0 such that the mapping is τ_0-Lipschitz continuous in $U(x^+, \delta_0)$, i.e.

$$\|f(x) - f(y)\| \leq \tau_0 \|x - y\|, \qquad x, y \in U(x^+, \delta_0) . \tag{13.13}$$

(d) There exist two positive scalars τ_1 and $\alpha \leq 1$ such that

$$\|A(x) - A(y)\| \leq \tau_1 \|x - y\|^\alpha , \qquad x, y \in U(x^+, \delta_0) . \tag{13.14}$$

We shall make use of the following lemma, due to Gay (1975a) and whose proof is omitted.

Lemma 13.1

If conditions (a)–(d) are satisfied, then for every scalar $\beta \in (0, 1)$ there exists a scalar $\delta_1 \in (0, \delta_0]$ such that, for every $z \in U(x^+, \delta_1)$, the following inequality is true:

$$\|z - x^+\| \leq \frac{\|A^{-1}(x^+)\| \, \|f(z)\|}{1 - \beta} . \tag{13.15}$$

Theorem 13.1

Consider the sequence x_i where x_1 is given and x_i is generated through the cycle defined by ALGORITHM 15. Assume that conditions (a)–(d) hold and that, for all cycles and for $k = 1, \ldots, r(i)$, P_k and V_k satisfy

$$\|P_k(V_k^{\mathrm{T}}A(u_k)P_k)^{-1}V_k^{\mathrm{T}}\| \le \tau_2 , \tag{13.16}$$

$$\mathrm{Cond}(V_1, \ldots, V_{r(i)}] \le \tau_3 , \tag{13.17}$$

for some scalars $\tau_2 > 0$, $\tau_3 \ge 1$. Then there exists a $\delta \in (0, \delta_0]$ such that for any $x_1 \in U(x^+, \delta^+)$ the sequence x_i converges to x^+ with Q-order no less than $1 + \alpha$.

Proof
Define τ_4 by

$$\tau_4 = (1 + \tau_0\tau_2)^n \tag{13.18}$$

and let δ_2 satisfy $\tau_4\delta_2 \le \delta_0$. Then for any $y_1 \in U(x^+, \delta_2)$ and $k = 1, \ldots, r(i)$ it follows that $y_{k+1} \in U(x^+, \delta_2)$ and

$$\|y_{k+1} - x^+\| \le \tau_4\|y_1 - x^+\| . \tag{13.19}$$

Inequality (13.19) follows from (13.10), (13.9), $f(x^+) = 0$, condition (c), (13.16) and the inequality

$$\|y_{k+1} - x^+\| \le (1 + \tau_0\tau_2)\|y_k - x^+\| . \tag{13.20}$$

By elementary calculations we have now, with $r = r(i) + 1$,

$$\|f(y_r)\| \le \|(V^{\mathrm{T}})^{-1}\| \sum_{k=1}^{r(i)} \|V_k^{\mathrm{T}}f(y_r)\| . \tag{13.21}$$

Consider now the expansion of $f(y_r)$ around y_k, $k = 1, \ldots, r(i)$:

$$f(y_r) = f(y_k) + \overline{A}_k(y_r - y_k) , \tag{13.22}$$

where, with $\theta_{kj} \in (0, 1)$, $k, j = 1, \ldots, r(i)$,

$$\overline{A}_k = \begin{bmatrix} a_1^{\mathrm{T}}[y_k + \theta_{k1}(y_r - y_k)] \\ \cdot \\ a_n^{\mathrm{T}}[y_k + \theta_{kn}(y_r - y_k)] \end{bmatrix}. \tag{13.23}$$

Then elementary calculation gives

$$V_k^T f(y_r) = V_k^T[f(y_k) + A(u_k)(y_r - y_k)] + V_k^T[\bar{A}_k - A(u_k)](y_r - y_k) \ .$$
$$(13.24)$$

Using the relation $H_p A^T(u_k)V_k = 0$ for $k < p$, see (7.50), we have $V_k^T[f(y_k) +$

$$A(u_k)(y_r - y_k)] = V_k^T f(y_k) - V_k^T A(u_k) \sum_{p=k}^{r(i)} P_p d_p = V_k^T f(y_k) - V_k^T A(u_k)P_k d_k = 0.$$

Hence

$$V_k^T f(y_r) = V_k^T[\bar{A}_k - A(u_k)](y_r - y_k) \ . \tag{13.25}$$

From (13.25) it follows that $\|V_k^T f(y_r)\| \leqslant \|V_k^T\| \ \|\bar{A}_k - A(u_k)\| \ \|y_r - y_k\| \leqslant 2n\tau_1\|V^T\|(\max\|y_p - y_q\|)^\alpha\|y_r - y_k\|$, where max is over p, $q = 1, \ldots, r(i)$ and $V = (V_1, \ldots, V_{r(i)})$. From (13.19) we have

$$\|V_k^T f(y_r)\| \leqslant \tau_5\|V^T\| \ \|y_1 - x^+\|^{1+\alpha}, \tag{13.26}$$

with τ_5 given by

$$\tau_5 = \tau_1\tau_2^{1+\alpha}n^{1/2}2^{1+2\alpha} \ . \tag{13.27}$$

Hence

$$\|f(y_r)\| \leqslant n\tau_5 \ \text{Cond}(V^T)\|y_1 - x^+\|^{1+\alpha} \ . \tag{13.28}$$

Using now Lemma 13.1 with $z = x_{i+1} = y_r$ and assuming that $x_i \in U(x^+, \delta_1) \cap U(x^+, \delta_2)$, (13.28) gives

$$\|x_{i+1} - x^+\| \leqslant \tau_6\|x_i - x^+\|^{1+\alpha} \ , \tag{13.29}$$

with τ_6 defined by

$$\tau_6 = \frac{n\tau_5\text{Cond}(V^T)\|A^{-1}(x^+)\|}{1 - \beta} \ . \tag{13.30}$$

Local convergence with Q-order no less than $1 + \alpha$ follows then from (13.30) if δ^+ is chosen so that the following two inequalities are true:

$$\delta^+ \leq \min(\delta_0, \delta_1, \delta_2) \ , \tag{13.31}$$

$$\delta_6(\delta^+)^{1+\alpha} \leq 1/2 \ , \tag{13.32}$$

Q.E.D.

Remark 13.2
Usually α has the value one, implying that the rate of convergence is Q-quadratic.

13.4 SPECIAL CASES OF THE SCALED BLOCK NONLINEAR ABS ALGORITHM

First we observe that the Newton method is a special case of the nonlinear ABS algorithm, obtained when $r(i) = 1$ for all i. Indeed in such a case there is only one block of size n by n and $y_2 = x_i - d_1 = y_1 - A(y_1)^{-1}f(y_1)$ is the Newton iteration.
    If $r(i) = n$ for all i then ALGORITHM 15 can be written as follows.

ALGORITHM 16: The Scaled Equationwise Nonlinear ABS Cycle
(A16) Let $y_1 = x_i$ and $H_1 = I$. Set $k = 1$.
(B16) Define $u_k \in R^n$ as in step (B15) and compute $A_k = A(u_k)$.
(C16) Define the search vector p_k by

$$p_k = H_k^T z_k \ , \tag{13.33}$$

where $z_k \in R^n$ is arbitrary, save that

$$p_k^T A_k^T v_k \neq 0 \ . \tag{13.34}$$

(D16) Define the step size α_k by

$$\alpha_k = \frac{v_k^T f(y_k)}{p_k^T A_k^T v_k} \ . \tag{13.35}$$

(E16) Update the minor iterate y_k by

$$y_{k+1} = y_k - \alpha_k p_k \ . \tag{13.36}$$

    If $k = n$, set $x_{i+1} = y_{n+1}$, the cycle is completed.
(F16) Update H_k by

$$H_{k+1} = H_k - H_k A_k^T v_k w_k^T H_k \ , \tag{13.37}$$

where $w_k \in R^n$ is arbitrary, save that

$$w_k{}^T H_k A_k{}^T v_k = 1 .$$ (13.38)

(G16) Increment the index k by one and go to (B16).

ALGORITHM 16 is a generalization of the nonlinear ABS algorithm considered by Abaffy *et al.* (1987a), where v_k and u_k were limited to the choices $v_k = e_k$ and $u_k = y_k$. The algorithms of Brown (1969) and of Brent (1973) are obtained in such a case by the choices $z_k = e_k$, $w_k = e_k/e_k{}^T H_k a_k(y_k)$ and $z_k = a_k(y_k)$, $w_k = z_k/z_k{}^T H_k z_k$. Note that a reformulation of the Brown method in terms of a nonlinear escalator method is due to Gergely (1980). Generalizations of the Brown and Brent methods, with local convergence properties of the type given by us, are presented by Gay (1975a) and Cosnard (1975). The cycle in the Cosnard class defines y_{k+1} as any vector in the linear variety containing all solutions of the k equations $a_j(y_j)^T y_{k+1} = f_j(y_j)$, $j = 1$, ..., k. Since by proper selection of z_k any such a vector can be generated by ALGORITHM 16, the Cosnard class and ALGORITHM 16 with $v_k = e_k$ are equivalent. The algorithm of Gay is equivalent (Abaffy *et al.* 1987a) to that subclass of ALGORITHM 16 where $v_k = e_k$, H_k is symmetric and z_k is chosen so that $|p_k{}^T a_k(y_k)| \geqslant \beta \|H_k a_k(y_k)\| \|p_k\|$, with $\beta \in (0, 1]$ and p_k is an eigenvector of H_k.

 We now discuss some choices of the scaling and the search vectors that satisfy conditions (13.16) and (13.17). First we consider the case $r(i) = n$. Condition (13.16) reads then

$$\left\| \frac{p_k v_k{}^T}{v_k{}^T A_k p_k} \right\| \leqslant \tau_2 .$$ (13.39)

Condition (13.39) can be satisfied by choosing z_k so that the angle between p_k and $A_k{}^T v_k$ or $A_k p_k$ and v_k is uniformly less than $\pi/2$, for all k and i, as established by the following theorem.

Theorem 13.2
Let $\beta \in [0, \pi/2]$ be arbitrary but fixed for all k and i and assume that either one of the following conditions hold:

$$|(p_k, A_k{}^T v_k) \measuredangle| \leqslant \beta$$ (13.40)

or

$$|(A_k p_k, v_k) \measuredangle| \leqslant \beta .$$ (13.41)

Assume that conditions (a)–(d) are satisfied in $U(x^+, \delta)$. Then there exists a $U(x^+, \delta)$ with $\bar{\delta} \leqslant \delta$ such that condition (13.39) is satisfied for a suitable τ_2 for all $u_k \in U(x^+, \bar{\delta})$.

Proof

From conditions (a)–(d) it follows easily that there exists a $\bar{\tau} \leqslant \|A(x^+)^{-1}\|$ and a $\bar{\delta} \leqslant \delta$ such that, for $x \in U(x^+, \delta)$, $A(x)$ is invertible and the following inequality is true:

$$\|A(x)^{-1}\| < \bar{\tau} \ . \tag{13.42}$$

Suppose now that (13.40) holds. Then we can write

$$\left\| \frac{p_k v_k^{\mathrm{T}}}{v_k^{\mathrm{T}} A_k p_k} \right\| \leqslant \frac{\|v_k\|}{\cos\beta \|A_k^{\mathrm{T}} v_k\|} \ . \tag{13.43}$$

If U is a nonsingular matrix, the identity $U^{-1}Uu = u$ gives

$$\|Uu\| \geqslant \frac{\|u\|}{\|U^{-1}\|} \ . \tag{13.44}$$

Assuming now that $u_k \in U(x^+, \bar{\delta})$ and using (13.44) with $u = v_k$, $U = A_k^{\mathrm{T}}$, (13.43) gives

$$\left\| \frac{p_k v_k^{\mathrm{T}}}{v_k^{\mathrm{T}} A_k p_k} \right\| \leqslant \frac{\|A_k^{-1}\|}{\cos\beta} \ . \tag{13.45}$$

Taking into account (13.42), it follows that (13.39) is satisfied with $\bar{\tau}_2 = \bar{\tau}/\cos\beta$ for all $u_k \in U(x^+, \bar{\delta})$. If (13.41) holds, then the inequality corresponding to (13.43) is

$$\left\| \frac{p_k v_k^{\mathrm{T}}}{v_k^{\mathrm{T}} A_k p_k} \right\| \leqslant \frac{\|p_k\|}{\cos\beta \|A_k p_k\|} \tag{13.46}$$

and again (13.45) follows by using (13.44) with $u = p_k$, $U = A_k$. Q.E.D.

Remark 13.3

If we set $v_k = A_k^{-\mathrm{T}} p_k$, then condition (13.40) is satisfied with $\beta = 0$. If moreover we set $u_k = x_i$ for $k = 1, \ldots, n$, then the scaling and the search vectors have the properties pertaining to the optimally stable subclass. Since the search vectors are then orthogonal, condition (13.17) is clearly satisfied with τ_3 a suitable multiple of τ. The nonlinear versions of the Huang, the Fridman and the Craig algorithms with $u_k = x_i$ are therefore locally Q-superlinearly (usually Q-quadratic) convergent.

Remark 13.4

If we set $v_k = A_k p_k$, then condition (13.41) is satisfied with $\beta = 0$. If moreover we set $u_k = x_i$ for $k = 1, \ldots, n$, then the scaling and the search vectors have the properties pertaining to the orthogonally scaled subclass. Since the scaling vectors are in such a case orthogonal, condition (13.17) is clearly satisfied. The nonlinear versions of the implicit QR and the minimum A^TA-iterations conjugate gradient-type algorithms with $u_k = x_i$ are therefore locally Q-superlinearly convergent.

Another sufficient condition for the validity of (13.39) is given by the following theorem; see Abaffy and Galantai (1986) for the proof.

Theorem 13.3
Suppose that, for $k = 1, \ldots, n$ and all i the norm of H_k is bounded by a constant $\bar{\tau} > 1$ and that the angle between z_k and $H_kA_k{}^Tv_k$ satisfies

$$|(z_k, H_kA_k{}^Tv_k) \not\prec | \leqslant \beta , \qquad \beta \in \left[0, \frac{\pi}{2}\right) . \tag{13.47}$$

Then condition (13.39) is satisfied with $\tau_2 = \bar{\tau} \cos \beta$.

Remark 13.5
Conditions for the matrices H_k to be bounded in norm are given in Chapter 3, including the case when H_k is symmetric. Condition (13.47) is satisfied with $\beta = 0$ if $z_k = H_kA_k{}^Tv_k$. Therefore all nonlinear modified scaled Huang algorithms are locally Q-superlinearly convergent.

We consider now the case when $r(i) < n$, so that there is at least one block containing more than one equation. Let us assume that $u_k = x_i$. Suppose that Z_k is chosen so that the matrices P_k, V_k are block biorthogonal ,say

$$P^TA^T(x_i)V = C , \tag{13.48}$$

where $P = (P_1, \ldots, P_{r(i)})$, $V = (V_1, \ldots, V_{r(i)})$ and C is block diagonal, $C = \text{diag}(C_1, \ldots, C_{r(i)})$. Assume that $\text{Cond}(C) \leqslant \tau$. Considering the left-hand side in (13.16) and using (13.48), we have

$$\|P_k[V_k{}^TA(x_i)P_k]^{-1}V_k{}^T\| \leqslant \|P\| \|C_k^{-1}\| \|V\| . \tag{13.49}$$

From (13.48) we also have $\|P\| \leqslant \|C\| \|V^{-1}\| \|A^{-1}(x_i)\|$, hence

$$\|P_k[V_k{}^TA(x_i)P_k]^{-1}V_k{}^T\| \leqslant \tau \, \text{Cond}(V) \|A^{-1}(x_i)\| . \tag{13.50}$$

Biorthogonal blocks with C_k equal to a unit matrix can be constructed if Z_k is taken such that $\text{Range}(Z_k)$ lies in the orthogonal complement to $\text{Span}[H_kA^T(x_i)V_{k+1}, \ldots, H_kA^T(x_i)V_{r(i)}]$. If $\text{Cond}(V)$ is bounded, it follows from the continuity of the Jacobian in x^+ that (13.16) is satisfied in a neighbourhood of x^+.

Still assuming that $u_k = x_i$, the matrix $V^TA(x_i)P$ is block lower triangular. Since its inverse is also block lower triangular, with blocks $[V_k{}^TA(x_i)P_k]^{-1}$ on the main diagonal, $k = 1, \ldots, r(i)$, we have the inequality

$$\|(V_k^T A(x_i) P_k)^{-1}\| \leqslant \|(V^T A(x_i) P)^{-1}\| \; , \tag{13.51}$$

implying that

$$\|P_k[V_k^T A(x_i) P_k]^{-1} V_k^T\| \leqslant \text{Cond}(P) \, \text{Cond}(V) \|A(x_i)^{-1}\| \; . \tag{13.52}$$

13.5 ANOTHER Q-SUPERLINEAR CONVERGENCE THEOREM

The following Theorem simplifies the proof of Theorem 13.1 for certain choices of the blocks P_k and U_k.

Theorem 13.4
Assume that conditions (a), (c) and (d) hold, that $A(x)$ is nonsingular, $\|A^{-1}(x)\| \leqslant \tau_3$, if $x \in U(x^+, \delta_0)$ and that, for all cycles, P_k, V_k satisfy either

$$P_k = A(u_k)^T V_k \tag{13.53}$$

or

$$V_k = A(u_k) P_k \; . \tag{13.54}$$

Assume moreover that with $V = (V_1, \ldots, V_{r(i)})$.

$$\text{cond}(V) \leqslant \tau_2 \; , \qquad \tau_2 \geqslant 1 \; . \tag{13.55}$$

Then there exists a number $\delta^+ > 0, 0 < \delta^+ \leqslant \delta_0$, such that for any $x_0 \in U(x^+, \delta^+)$ the sequence x_i converges to x_+ with Q-order no less than $1 + \alpha$.

Proof
Using the singular value decomposition theorem, it is easy to show that for arbitrary $Q \in R^{n,k}$, with $k \leqslant n$, and rank$(Q) = k$, the following equation is true:

$$Q(Q^T Q)^{-1} Q^T = \bar{I}_k \; , \tag{13.56}$$

where $\bar{I}_k \in R^{n,n}$ is a diagonal matrix with k units on the main diagonal. Let

$$\tau_4 = (1 + \tau_0 \tau_3)^n, \tag{13.57}$$

and let θ_1 be such that

$$\tau_4 \theta_1 \leqslant \delta_0 \; . \tag{13.58}$$

Then, for any $y_0 \in U(x^+, \delta_1)$, it follows that $y_k \in U(x^+, \delta_0)$ and, from $f(x^+) = 0$, (13.9), (13.10) and (13.56),

$$\|y_{k+1} - x^+\| \leq \|y_k - x^+\|$$
$$+ \|Q_k(Q_k^TQ_k)^{-1}Q_k\| \ \|A(u_k)^{-1}\| \ \|f(y_k) - f(x^+)\| \ ,$$
$$(13.59)$$

where

$$Q_k = P_k \quad \text{or} \quad Q_k = A(u_k)P_k \ , \tag{13.60}$$

depending on whether (13.53) or (13.54) holds. Thus, using (13.13), we get

$$\|y_k - x^+\| \leq \tau_4 \|y_1 - x^+\| \ . \tag{13.61}$$

Also we have

$$\|f(y_r)\| \leq \|V^{-1}\| \sum_{k=1}^{r} \|V_k^T f(y_r)\| \ . \tag{13.62}$$

Taking the Taylor expansion of $f(y_r)$ around y_k, we obtain

$$f(y_r) = f(y_k) + \overline{A}_k(y_r - y_k) \ , \tag{13.63}$$

where

$$\overline{A}_k = \begin{bmatrix} a_1^T[y_k + \theta_1(y_r - y_k)] \\ . \\ a_n^T[y_k + \theta_n(y_r - y_k)] \end{bmatrix} \tag{13.64}$$

and $\theta_k \in (0, 1)$, $i = 1, \ldots, n$. Observe that

$$V_k^T f(y_r) = V_k^T f(y_r) + V_k^T A(u_k)(y_r - y_k)$$
$$+ V_k^T[\overline{A}_k - A(u_k)](y_r - y_k) \ . \tag{13.65}$$

However, $V_k^T f(y_k) + V_k^T A(u_k)(y_r - y_k) = V_k^T f(y_k) - V_k^T A(u_k) \sum_{i=k}^{r(i)} P_i d_i = 0$ because $H_i A(u_k)^T V_k = 0$ for all $i > k$. Therefore we get

$$V_k^{\mathrm{T}}f(y_r) = V_k^{\mathrm{T}}[\overline{A}_k - A(u_k)](y_r - y_k) \ . \tag{13.66}$$

It follows from the foregoing that, using (13.14), (13.61) and (13.62), we have $\|f(y_r)\| \leqslant \mathrm{Cond}(V)\tau_1(2 \max\|y_i - y_j\|, \ i, j = 1, \ldots, r)^{\alpha}\|y_r - y_k\| \leqslant \tau_5 \, \mathrm{Cond}(V)\|y_1 - x^+\|^{1+\alpha}$, with a suitable τ_5. Using Lemma 13.1 with $z = x_{i+1} = y_r$, it follows that for every $0 < \beta < 1$ and assuming that $x_i \in U(x^+, \delta_2) \cap U(x^+, \delta_1)$, we have

$$\|x_{i+1} - x^+\| \leqslant \tau_6\|x_i - x^+\|^{1+\alpha} \ , \tag{13.67}$$

with

$$\tau_6 = \frac{\tau_5 \, \mathrm{Cond}(V)\|A(x^+)^{-1}\|}{1 - \beta} \ . \tag{13.68}$$

If δ^+ is chosen such that

$$\delta^+ \leqslant \min(\delta_0, \, \delta_1, \, \delta_2) \tag{13.69}$$

and

$$\tau_6(\delta^+)1 + \alpha < 1/2 \ , \tag{13.70}$$

then the theorem is true. Q.E.D.

Remark 13.6
It is possible to avoid the explicit use of the matrix H_k in the proof if we suppose that the matrices $A(u_k)^{\mathrm{T}}V_k$ are mutually orthogonal, i.e.

$$V_k^{\mathrm{T}}A(u_k)A(u_j)^{\mathrm{T}}V_j = 0 \ , \qquad k \neq j \ . \tag{13.71}$$

In this case, (13.66) is valid again and the theorem is true, having only used the block biorthogonal property of V_k.

13.6 NUMERICAL EXPERIMENTS

A limited number of numerical experiments with nonlinear ABS algorithms have been presented by Spedicato and Bodon (1987) and Bodon (1987). Two particular cases of ALGORITHM 16 are considered, say the nonlinear modified Huang algorithm, equivalent to the Brent method, given by

$$v_k = e_k, \ u_k = y_k, \ z_k = H_k a_k(y_k), \ w_k = \frac{z_k}{z_k^{\mathrm{T}}z_k} \ , \tag{13.72}$$

and the nonlinear implicit LU algorithm, equivalent to the Brown method, with pivoting of equations, given by

$$v_k = e_k, \ u_k = y_k, \ z_k = e_k, \ w_k = \frac{e_k}{e_k^T H_k a_k(y_k)} \ , \tag{13.73}$$

where the kth equation is defined by a pivoting strategy based upon maximizing with respect to j the quantity $|e_k^T H_k a_j(y_k)|, j = k, \ldots, n$. The two algorithms have been implemented in FORTRAN and run in single precision on a Burroughs 6800 (about 12 decimal digits). The linear dependence test $H_k a_k(y_k) = 0$, implying that $y_{k+1} = y_k$, has been implemented as the test

$$\|H_k a_k(y_k)\| \leqslant 10^{-6} \|a_k(y_k)\| \ . \tag{13.74}$$

The iteration is terminated as soon as one of the following cases occurs.

— $\|f(x_i)\| \leqslant 10^{-10}$.
— $\|x_{i+1} - x_i\| \leqslant 10^{-10} \|x_{i+1}\|$.
— $\|f(x_i)\|$ does not decrease in three successive iterations.

The codes have been tested on about 30 problems defined by Cosnard (1975) and Spedicato (1977), with a standard starting point x_1 and four additional starting points τx_1 ($\tau = 1.1$, 10 and 100). For comparison purposes, the sophisticated code due to More and Cosnard (1979), based upon the Brent method, has been run on the same problems. Some of the results are reproduced in Table 13.1, giving the number of

Table 13.1 — Results on test problems

Problem	n	Method	IT	$\|f\|, \tau = 1$	IT	$\|f\|, \tau = 100$
Powell singular	4	B	21	0.54E − 10	28	0.51E − 10
Powell singular	4	MH	21	0.54E − 10	28	0.51E − 10
Powell singular	4	LU	25	0.52E − 10	32	0.50E − 10
Schubert	10	B	4	0.29E − 10	12	0.29E − 10
Schubert	10	MH	4	0.42E − 10	12	0.29E − 10
Schubert	10	LU	4	0.53E − 10	11	0.14E − 10
Brown	10	B	5	0.92E − 10	21	0.80E − 10
Brown	10	MH	7	0.15E − 10	25	0.65E − 10
Brown	10	LU	7	0.44E − 10	91	0.83E − 10
Brown	40	B	10	0.77E − 9	9	0.41E − 9
Brown	40	MH	11	0.13E − 8	9	0.80E − 9
Brown	40	LU	8	0.50E − 9		Diverges
Trigonometric	4	B	26	0.82E − 10	31	0.71E − 10
Trigonometric	4	MH	25	0.73E − 10	29	0.13E − 10
Trigonometric	4	LU	41	0.70E − 10	42	0.50E − 10

iterations and the infinity norm of f in the computed solution. The symbols B, MH and LU refer to the Brent method, the modified Huang method and the implicit LU method.

The modified Huang algorithm, which is theoretically equivalent to the Brent algorithm, appears to be also numerically equivalent, giving usually the same number of iterations and very similar accuracy (in many cases the first five digits are identical). The implicit LU method tends to be less accurate, requires more iterations and fails more frequently. It is to be noted that the number of failures can be reduced by adopting a line search in the definition of x_{i+1}, say

$$x_{i+1} = x_i - \beta_i(y_{n+1} - x_i) \ , \tag{13.75}$$

where β_i is determined so that $\|f(x_{i+1})\| < \|f(x_i)\|$.

13.7 FINAL REMARKS

We have presented a very general class of algorithms for nonlinear systems, which includes the Newton, the Brown and the Brent methods. Our local convergence results apply, *inter alia*, to the nonlinear versions of the implicit LQ and QR algorithms and to the minimum AA^T and minimum A^TA-iterations conjugate gradient-type algorithms. Many questions have not been discussed in this chapter and require further investigation. We now list some of them.

— Extension of the local convergence results in the case when the Jacobian is not evaluated exactly but is estimated by finite differences or quasi-Newton techniques. This extension has been provided by Huang (1988a) for the restricted class considered by Abaffy *et al.* (1987a), assuming finite difference approximations.
— Analysis of the global convergence properties and, more generally, of the structure of the attraction regions.
— Study of the computational complexity of the class as a function of the block selections. See Schmidt and Hoyer (1978) and Hoyer (1981) for results on the Brown and the Brent methods.
— Analysis of the numerical stability of the class as a function of the parameter choices and the matrix formulation.
— Formulation of the class in a parallel framework.
— Effects of line searches in the definition of the major iterates; see for instance (13.75), and possibly also of the minor iterates (change (13.10) into $y_{k+1} = y_k - \beta_k P_k d_k$, with β_k chosen to reduce a suitable norm of f).
— Formulation of the class for problems where f has a special structure; for instance it represents Kuhn–Tucker conditions. See Abaffy *et al.* (1987b) and Huang (1988b) for preliminary results on the application of nonlinear ABS methods to unconstrained optimization and to nonlinear least squares.

13.8 BIBLIOGRAPHICAL REMARKS

The nonlinear ABS algorithm has been mainly developed by Abaffy and Galantai. The equationwise formulation has appeared in the paper by Abaffy *et al.* (1987a).

The blockwise formulation has appeared in the paper by Abaffy and Galantai (1986) (see also Galantai (1988, 1989)). Theorems 13.1 and 13.3 are due to Abaffy and Galantai (1986). Theorem 13.2 is a generalization of a theorem in the paper by Abaffy and Galantai (1986). Theorem 13.4 is due to Abaffy (1988a).

References

REFERENCES

Abaffy, J. (1979) A Lineáris Egyenletrendszerek Általános Megoldásának Egy Módszerosztálya, *Alkalmaz. Mat. Lapok*, **5**, 223–240.

Abaffy, J. (1984) A Generalization of the ABS Class, *Proceedings of the Second Conference on Automata Languages and Mathematical Systems*, University of Economics, Budapest, pp. 7–11.

Abaffy, J. (1986) Some Special Cases of the ABS Class for Band Matrices, *Proceedings of the Fourth Conference on Automata Languages and Mathematical Systems*, University of Economics, Budapest, pp. 9–15.

Abaffy, J. (1987a) *Error Analysis and Stability for Some Cases in the ABSg Class*, Report No. TR 190, NOC, Hatfield Polytechnic.

Abaffy, J. (1987b) *Preliminary Test Results with Some Algorithms of the ABSg Class*, Report No. TR 193, NOC, Hatfield Polytechnic.

Abaffy, J. (1987c) *Optimal Methods in Broyden's Stability Sense*, Report No. TR 194, NOC, Hatfield Polytechnic.

Abaffy, J. (1988a) A Superlinear Convergency Theorem in the ABSg Class for Nonlinear Algebraic Equations, *J. Optim. Theory Appl.*, **59**(1), 39–43.

Abaffy, J. (1988b) *Equivalence of a Generalization of Sloboda's Algorithm with a Subclass of the Generalized ABS Algorithm for Linear Systems*, Report No. DMSIA 1988/1, University of Bergamo.

Abaffy, J. (1988c) *Derivation of Lanczos Type Methods in the Generalized ABS Algorithm for Linear Systems*, Report No. DMSIA 1988/2, University of Bergamo.

Abaffy, J. (1988d) *Formulation of the Conjugate Gradient Type Algorithm of Fridman for Indefinite Symmetric Systems in the Subclass of the Generalized Symmetric ABS Algorithms*, Report No. DMSIA 1988/3, University of Bergamo.

Abaffy, J. (1988e) *Reorthogonalized Methods in the ABSg Class*, Report No. DMSIA 1988/17, University of Bergamo.

Abaffy, J., Broyden, C. G., and Spedicato, E. (1982) *A Class of Direct Methods for Linear Systems II*, Report No. SOFTMAT 1982/21, University of Bergamo.

Abaffy, J., Broyden, C. G., and Spedicato, E. (1983) *Numerical Performance of the Pseudosymmetric Algorithm in the ABS Class*, Report No. SOFTMAT 1983/80, University of Bergamo.

Abaffy, J., Broyden, C. G., and Spedicato, E. (1984a) *A Class of Direct Methods of Quasi-Newton Type for General Linear Systems*, Monografia SOFTMAT 33, IAC, Rome.

Abaffy, J., Broyden, C. G., and Spedicato, E. (1984b) A Class of Direct Methods for Linear Equations, *Numer. Math.*, **45**, 361–376.

Abaffy, J., and Dixon, L. C. W. (1987) *On Solving Sparse Band Systems with Three Algorithms of the ABS Family*, Report No. TR 191, NOC, Hatfield Polytechnic.

Abaffy, J. and Galantai, A. (1986) Conjugate Direction Methods for Linear and Nonlinear Systems of Algebraic Equations, *Colloq. Math. Soc. Janos Bolyai*, **50**, 481–502.

Abaffy, J., Galantai, A., and Spedicato, E. (1984c) *Convergence Properties of the ABS Algorithm for Nonlinear Algebraic Equations*, Report No. DMSIA 1984/7, University of Bergamo.

Abaffy, J., Galantai, A., and Spedicato, E. (1987a) The Local Convergence of ABS Methods for Nonlinear Algebraic Systems, *Numer. Math.*, **51**, 429–439.

Abaffy, J., Galantai, A., and Spedicato, E. (1987b) *Application of ABS Class to Unconstrained Function Minimization*, Report No. DMSIA 1987/14, University of Bergamo.

Abaffy, J., Galantai, A., and Spedicato, E. (1989) *Forward Error Analysis in ABS Methods*, Preprint, University of Bergamo.

Abaffy, J., and Spedicato, E. (1982) A Class of Direct Methods for Linear Equations I: Basic Properties, Report No. IAMI 82/4, IAMI, Milan.

Abaffy, J., and Spedicato, E. (1983a) *Computational Experience with a Class of Direct Methods for Linear Systems*, Report No. SOFTMAT 1983/1, University of Bergamo.

Abaffy, J., and Spedicato, E. (1983b) *On the Symmetric Algorithm in the ABS Class of Direct Methods for Linear Systems*, Report No. SOFTMAT 1983/7, University of Bergamo.

Abaffy, J., and Spedicato, E. (1984) A Generalization of Huang's Method for Solving Systems of Linear Algebraic Equations, *Boll. Unione Mat. Ital.*, **63-B**, 517–529.

Abaffy, J., and Spedicato, E. (1985) *A Generalization of the ABS Algorithm for Linear Systems*, Report No. DMSIA 1985/4, University of Bergamo.

Abaffy, J., and Spedicato, E. (1987) Numerical Experiments with the Symmetric Algorithm in the ABS Class for Linear Systems, *Optimization*, **18**(2), 197–212.

Abaffy, J., and Spedicato, E. (1988) *Orthogonally Scaled and Optimally Stable Algorithms in the Scaled ABS Class for Linear Systems*, Report No. DMSIA 1988/8, University of Bergamo.

Abaffy, J., and Spedicato, E. (1989) *On the Use of the ABS Algorithm for some Linear Programming Problems*, Preprint, University of Bergamo.

Albert, A. (1972) *Regression and the Moore–Penrose Pseudoinverse*, Academic

Press, New York.

Bertocchi, M., and Spedicato, E. (1988a) *Computational Performance on the IBM 3090 of the Modified Huang and the Implicit Gauss–Cholesky Algorithm Versus the Gaussian Solver in the ESSL Library on Ill-Conditioned Problems*, Report No. DMSIA 1988/4, University of Bergamo.

Bertocchi, M., and Spedicato, E. (1988b) *Vectorizing the Implicit Gauss–Cholesky Algorithm of the ABS Class on the IBM 3090 VF*, Report No. DMSIA 1988/22, University of Bergamo.

Bertocchi, M., and Spedicato, E. (1988c) *Vectorizing the Modified Huang Algorithm of the ABS Class on the IBM 3090*, Report No. DMSIA 1988/23, University of Bergamo.

Björck, Å. (1967) Solving Linear Least Squares Problems by Gram–Schmidt Orthogonalization, *BIT, Nord. Tidskr. Infbehandl.*, **7**, 1–21.

Björck, Å. (1986) *Least Squares Methods*, Preprint, Mathematical Department, University of Linköping.

Bodon, E. (1987) *Algoritmi ABS per minimi quadrati ed equazioni nonlineari*, PhD Dissertation, University of Genova.

Bodon, E. (1988) *Globally Optimally Conditioned Updates in the ABS Class for Linear Systems*, Report No. SOFTMAT 1988/1, University of Bergamo.

Bodon, E. (1989) *Biconjugate Algorithms in the ABS Class I: Alternative Formulation and Theoretical Properties*, Report DMSIA 1989/3, University of Bergamo.

Boullion, T., and Odell, P. (1971) *Generalized Inverse Matrices*, Wiley, New York.

Brent, R. P. (1973) Some Efficient Algorithms for Solving Systems of Nonlinear Equations, *SIAM J. Numer. Anal.*, **10**, 324–344.

Brown, K. M. (1969) A Quadratically Convergent Newton-like Method Based Upon Gaussian Elimination, *SIAM J. Numer. Anal.*, **6**, 560–569.

Broyden, C. G. (1965) A Class of Methods for Solving Nonlinear Simultaneous Equations, *Math. Comput.*, **19**, 577–593.

Broyden, C. G. (1974) Error Propagation in Numerical Processes, *J. Inst. Math. Appl.*, **14**, 131–140.

Broyden, C. G. (1975) *Basic Matrices: an Introduction to Matrix Theory and Practice*, McMillan, London.

Broyden, C. G. (1985) On the Numerical Stability of Huang and Related Methods, *J. Optim. Theory Appl.*, **47**, 401–412.

Broyden, C. G. (1989) Error perturbation of the eigenvalues of the matrices generated by the Huang and the modified Huang algorithms, Report No. DMSIA 89/14, University of Bergamo.

Chandra, R. (1978) *Conjugate Gradient Methods for Partial Differential Equations*, Report No. 129, Yale University.

Cimmino, G. (1938) Calcolo Approssimato per le Soluzioni di Equazioni Lineari, *Ric. Sci. II*, **9**, 326–333.

Cline, R. E. (1964) Representations for the Generalized Inverse of a Partitioned Matrix, *SIAM J. Appl. Math.*, **12**, 588–600.

Cline, R. E., and Plemmons, R. J. (1976) L_2-Solutions to Undetermined Linear Systems, *SIAM Rev.*, **18**, 92–106.

Cosnard, M. Y. (1975) *Sur Quelques Methodes "Newton Like" de Resolution de Systèmes d'Equations Non Linéaires*, Report No. T.R. 239, University of Grenoble.

Craig, E. J. (1955) The N-Step Iterations Procedures, *J. Math. Phys.*, **34**, 64–73.

Dahlquist, G., and Björck, Å. (1974) *Numerical Methods*, Prentice Hall, Englewood Cliffs, NJ.

Daniel, J., Gragg, W. B., Kaufman, L., and Stewart, G. W. (1976) Reorthogonalization and Stable Algorithms for Updating the Gram–Schmidt QR Factorization, *Math. Comput.*, **30**, 772–795.

Deng, N. Y. (1987) *Nested Dissection Methods for Sparse Positive Definite Linear Systems of Equations*, Report No. TR. 197, NOC, Hatfield Polytechnic.

Deng, N. Y., and Spedicato, E. (1988) *Optimal Conditioning Parameter Selection in the ABS Class through a Rank Two Update Formulation*, Report No. DMSIA 1988/18, University of Bergamo.

Deng, N. Y., and Xiao Y. (1989) *The Minimum Norm Correction Class for Linear System of Equations*, Preprint, Department of Applied Mathematics, Beijing Polytechnic University.

Dennis, J. E., Jr., and Moré, J. J. (1977) Quasi-Newton Methods, Motivation and Theory, *SIAM Rev.*, **19**, 46–89.

Dennis, J. E., Jr., and Schnabel, R. B. (1983) *Numerical Methods for Nonlinear Equations and Unconstrained Optimization*, Prentice-Hall, Englewood Cliffs, NJ.

Elman, H. C. (1982) *Iterative Methods for Large Sparse Nonsymmetric Systems of Linear Equations*, Report No. 229, Computer Science Department, Yale University.

Egervary, E. (1960) On Rank-Diminishing Operations and their Applications to the Solution of Linear Equations, *Z. Angew. Math. Phys.*, **11**, 376–386.

Flachs, J. (1982) *On the Generalization of Updates for Quasi-Newton Methods*, Report No. NRIMS/W/766, Pretoria.

Fletcher, R. (1975) *Conjugate Gradient Methods for Indefinite Systems, Proceedings of the Dundee Conference on Numerical Analysis*, Springer, Berlin.

Forsythe, G. E., and Moler, C. B. (1967) *Computer Solution of Linear Algebraic Systems*, Prentice-Hall, Englewood Cliffs, NJ.

Fox, L., Huskey, H. D., and Wilkinson, J. H. (1948) Notes on the Solution of Algebraic Simultaneous Equations, *Quart. J. Mech. Appl. Math.*, **1**, 149–173.

Fragnelli, V., and Resta, G. (1985) *An ABS Method for Solving Suitable Structured Linear Systems is Supported by Parallel Architectures*, Preprint, Mathematical Institute, University of Genova.

Freund, R. (1983) *Über einige cg-ähnliche Verfahren zur Lösung Linearer Gleichungssyteme*, Dissertation, University of Würzburg.

Fridman, V. M. (1963) The Method of Minimum Iterations with Minimum Errors for a System of Linear Algebraic Equations with a Symmetrical Matrix, *USSR Comput. Math. Math. Phys.*, **2**, 362–362.

Galantai, A. (1987) *A Study of Error Propagation for the ABS Method*, Report No. DMSIA 1987/17, University of Bergamo.

Galantai, A. (1988) Block ABS Methods for Nonlinear Systems of Algebraic Equations, *Proceedings of the Workshop on Algorithms for Nonlinear Systems, Fort Collins*.

Galantai, A. (1989) A new convergence theorem for nonlinear ABS methods, Report N. DMSIA 1989/13, University of Bergamo.

212 References

Gauss, K. F. (1821) *Theoria Combinationis Observationum Erroribus Minimis Obnoxiae* (also in (1973) *Werke*, Vol. IV, Georg Olms, Hildesheim).

Gay, D. M. (1975a) *Brown's Method and Some Generalizations with Applications to Minimization Problems*, Report No. TR 75-225, Cornell University.

Gay, D. M. (1975b) *Implementing Brown's Method*, Report No. CNA-109, University of Texas at Austin.

Gerber, R., and Luk, F. (1981) *A Generalized Broyden Method for Solving Simultaneous Linear Equations*, SIAM J. Numer. Anal., **18**, 882–890.

George, J. A. (1973) Nested Dissection of a Regular Finite Element Mesh, *SIAM J. Numer. Anal.*, **10**, 345–363.

George, J. A., and Liu, J. W. (1981) *Computer Solution of Large Sparse Positive Definite Systems*, Prentice-Hall, Englewood Cliffs, NJ.

Gergely, J. (1980) Matrix Inversion and Solution of Linear and Nonlinear Systems by the Method of Bordering, Numerical Methods, *Colloq. Math. Soc. Janos. Bolyai*, **22**.

Golub, G. H., and Van Loan, C. F. (1983) *Matrix Computations*, Johns Hopkins University Press, Baltimore, MD.

Graham, A. (1979) *Matrix Theory and Applications for Engineers and Mathematicians*, Ellis Horwood, Chichester, West Sussex.

Greenstadt, J. (1970) Variations on variable metric methods, *Mathematics of Computation* **24**, 1—18.

Gu, S. (1988) *The Simplified ABS Algorithm*, Report No. RABSCA 1, Department of Applied Mathematics, Dalian University of Technology.

Hanson, R. J., and Lawson, C. L. (1974) *Solving Least Squares Problems*, Prentice-Hall, Englewood Cliffs, NJ.

Hegedüs, Cs. J., and Bodocs, L. (1982) *General Recursions for A-Conjugate Vector Pairs*, Report No. 1982/56, Central Research Institute for Physics, Budapest.

Hestenes, M. R. (1980) *Conjugate Direction Methods in Optimization*, Springer, Berlin.

Hestenes, M. R., and Stiefel, E. (1952) Methods of Conjugate Gradients for Solving Linear Systems, *J. Res. Natl. Bur. Stand.*, **49**, 409–436.

Householder, A. S. (1955) Terminating and Nonterminating Iterations for Solving Linear Systems, *J. Soc. Indust. Appl. Math.*, **3**, 67–72.

Householder, A. S. (1964) *The Theory of Matrices in Numerical Analysis*, Blaisdell.

Hoyer, W. (1981) Zur Effektivität Mehrstufiger Brown–Brent-Verfahren, *Beitr. Numer. Math.*, **10**, 57–69.

Huang, H. Y. (1975) A Direct Method for the General Solution of a System of Linear Equations, *J. Optim. Theory Appl.*, **16**, 429–445.

Huang, Z. (1988a) *A New Approach for Solving Nonlinear Least Squares Problems*, Preprint, Department of Statistics and Operations Research, Fudan University, Shanghai.

Huang, Z. (1988b) *Modified ABS Methods for Nonlinear Systems without Evaluating Jacobian Matrices*, Preprint, Department of Statistics and Operations Research, Fudan University, Shanghai.

Kaczmarz, S. (1937) Angenäherte Auflösung von Systemen Linearen Gleichungen, *Bull. Acad. Pol. Sci. Lett. A*, 355–357.

Lanczos, C. (1950) An Iteration Method for the Solution of the Eigenvalue Problem of Linear Differential and Integral Operators, *J. Res. Natl. Bur. Stand.*, **45**, 255–282.

Lanczos, C. (1952) Solution of Systems of Linear Equations by Minimized Iterations, *J. Res. Natl. Bur. Stand.*, **49**, 33–53.

Luenberger, D. G. (1973) *Introduction to Linear and Nonlinear Programming*, Addison-Wesley, Reading, MA.

Martin, R. S., and Wilkinson, J. H. (1968) Similarity Reduction of a General Matrix to Hessenberg Form, *Numer. Math.*, **12**, 349–368.

More, J. J., and Cosnard, M. Y. (1979) Numerical Solution of Nonlinear Equations, *ACM Trans.*, **5**, 64–85.

Morris, J. (1946) An Escalator Process for the Solution of Linear Simultaneous Equations, *Philos. Mag.*, **37**, 106–120.

Nashed, M. Z. (1976) *Generalized Inverses and Applications*, Academic Press, New York.

Noble, B., and Daniel, W. (1977) *Applied Linear Algebra*, Prentice Hall, Englewood Cliffs, NJ.

O'Leary, D. P. (1980) A Generalized Conjugate Gradient Algorithm for Solving a Class of Quadratic Programming Problems, *Linear Algebra Appl.*, **34**, 371–399.

Oprandi, M. (1987) *L'algoritmo LU Implicito dalla Classe ABS per Sistemi Lineari*, Dissertation, University of Bergamo.

Oren, S. S. (1972) *Self-Scaling Variable Metric Algorithms for Unconstrained Minimization*, PhD Dissertation, Stanford University.

Paige, C. C., and Saunders, M. A. (1975) Solution of Sparse Indefinite Systems of Linear Equations, *SIAM J. Numer. Anal.*, **12**, 617–629.

Parlett, B. N. (1980) *The Symmetric Eigenvalue Problem*, Prentice-Hall, Englewood Cliffs, NJ.

Phillips, C., and Cornelius, B. (1986) *Computational Numerical Methods*, Ellis Horwood, Chichester, West Sussex.

Phua, K. H. (1986) *Solving Sparse Linear Systems by ABS Methods*, Preprint No. 86/4, Computer Science Department, National University, Singapore.

Phua, K. H. (1988) Solving Sparse Linear Systems by an ABS Method that Corresponds to *LU* Decomposition, *BIT, Nord. Tidskr. Infbehandl.*, **28**, 709–718.

Pissanetzky, S. (1984) *Sparse Matrix Technology*, Academic Press, New York.

Pyle, L. D. (1964) Generalized Inverse Computations Using the Gradient Projection Method, *J. ACM*, **11**, 422–429.

Radicati di Brozolo, G., and Robert, Y. (1987) *Vector and Parallel CG-like Algorithms for Sparse non-Symmetric Systems*, Report No. ICE 10, IBM, Rome.

Rao, C. R., and Mitra, S. K. (1971) *Generalized Inverse of Matrices and its Applications*, Wiley, New York.

Rice, J. R. (1966) Experiments on Gram–Schmidt Orthogonalization, *Math. Comput.*, **20**, 325–328.

Schendel, V. (1984) *Introduction to Numerical Methods for Parallel Computers*, Ellis Horwood, Chichester, West Sussex.

Schmidt, J. W., and Hoyer, W. (1978) Die Verfahren vom Brown–Brent-Typ bei Gemischt Linearen-nichtlinearen Gleichungs-systemen, *Z. Angew. Math. Mech.*, **58**, 425–428.

Simon, H. D. (1982) *The Lanczos Algorithm for Solving Symmetric Linear Systems*, PhD Thesis, University of California, Berkeley.

Sloboda, F. (1978) A Parallel Projection Method for Linear Algebraic Systems, *Apl. Mat. Ceskosl. Akad. Ved.*, **23**, 185–198.

Sloboda, F. (1988) *A Projection Method of the Cimmino Type for Linear Algebraic Systems*, Report No. DMSIA 88/16, University of Bergamo.

Spedicato, E. (1976) Metodi Quasi-Newtoniani per Equazioni Algebriche e Minimizzazione Nonlineare: Sviluppi e Prospettive, *Boll. Unione Mat. Ital.*, **13-A**, 545–567.

Spedicato, E. (1977) *Algoritmi per la Soluzione di Equazioni Algebriche Nonlineari*, Report No. IAC III/29, Rome.

Spedicato, E. (1985) On the Solution of Linear Least Squares Through the ABS Class for Linear Systems, *Proceedings AIRO Conference, Tecnoprint, pp.* 89–98.

Spedicato, E. (1987a) Variationally Derived Algorithms in the ABS Class for Linear Systems, *Calcolo*, **24**, 241–246.

Spedicato, E. (1987b) *A Bound to the Condition Number of Positive Definite Matrices*, Report No. DMSIA 1987/1, University of Bergamo.

Spedicato, E. (1987c) *Optimal Conditioning Parameter Selection in the ABS Class for Linear Systems*, No. Report 203, Mathematische Institute, University of Würzburg.

Spedicato, E., and Bodon, E. (1987) *Numerical Experiments in the ABS Class for Nonlinear Systems of Algebraic Equations*, Report No. SOFTMAT 1987/1, University of Bergamo.

Spedicato, E., and Bodon, E. (1989a) *Solving Linear Least Squares by Orthogonal Factorization and Pseudoinverse Computation Via the Modified Huang Algorithm in the ABS Class*, Report No. DMSIA 1989/2, University of Bergamo.

Spedicato, E., and Bodon, E. (1989b) *Biconjugate Algorithms in the ABS Class II*: *Numerical Evaluation*, Report No. DMSIA 1989/4, University of Bergamo.

Spedicato, E., and Bodon, E. (1989c) *Solution of linear least squares via the ABS algorithm*, Report No. DMSIA 1989/5, University of Bergamo.

Spedicato, E., and Burmeister, W. (1988) A Strict Bound to the Condition Number of Bordered Positive Definite Matrices, *Computing*, **40**, 181–183.

Spedicato, E., and Greenstadt, J. (1978) On Some Classes of Variationally Derived Quasi-Newton Methods for Systems of Nonlinear Algebraic Equations, *Numer. Math.*, **29**, 363–380.

Spedicato, E., and Oren, S. S. (1976) Optimal Conditioning of Self-Scaling Variable Metric Algorithms, *Math. Program.*, **10**, 70–90.

Spedicato, E., and Vespucci, M. T. (1989) *Computational Performance of the Implicit Gram–Schmidt (Huang) Algorithm for Linear Algebraic Systems*, Report No. DMSIA 1989/6, University of Bergamo.

Stewart, G. W. (1973) Conjugate Direction Methods for Solving Systems of Linear Equations, *Numer. Math.*, **21**, 283–297.

Stoer, J. (1982) *Conjugate Directions Type Methods for Large Linear Systems*, in *Mathematical Programming, the State of Art*, Springer, Berlin.

Stoer, J., and Bulirsch, R. (1980) *Introduction to Numerical Analysis*, Springer, Berlin.

Stoer, J., and Freund, R. (1982) *On the Solution of Large Indefinite Systems of Linear Equations by Conjugate Gradient Algorithms*, Computing Methods in Applied Sciences and Engineering, Vol. V, North-Holland, Amsterdam.

Tanabe, K. (1971) Projection Method for Solving a Singular System of Linear Equations and its Applications, *Numer. Math.*, **17**, 203–214.

Voyevodin, V. V. (1983) *Linear Algebra*, Mir, Moscow.

Yang, Z. (1988a) *ABS Algorithms for Solving Systems of Indeterminate Equations*, Report No. RABSCA 2, Department of Applied Mathematics, Dalian University of Technology,

Yang, Z. (1988b) *The Properties of Iterative Points in ABS Algorithms and the Related Generalized Inverse Matrix*, Report No. RABSCA 3, Department of Applied Mathematics, Dalian University of Technology,

Yang, Z. (1988c) *An ABS Subclass with Optimal Conditioning Parameter Selection*, Report No. RABSCA 4, Department of Applied Mathematics, Dalian University of Technology.

Yang, Z. (1988d) On the numerical stability of the Huang and the modified Huang algorithms and related topics, Report No. RABSCA 5, Department of Applied Mathematics, Dalian University of Technology.

Zhao, J. (1981) Huang's Method for the Solution of Consistent Linear Equations and its Generalization, *Numer. Math.*, **3**, 8–17 (in Chinese).

Zeng, L. (1988) *An ABS Algorithm for Sparse Linear Systems*, Report No. RABSCA 5, Department of Applied Mathematics, Dalian University of Technology.

Zhu, M. (1987a) *The Implicit LL^T Algorithm for Sparse Nested Dissection Linear Systems*, Report No. TR 196, NOC, Hatfield Polytechnic.

Zhu, M. (1987b) *On the ABS Algorithms for Perturbated Linear Systems*, Report No. TR 198, NOC, Hatfield Polytechnic.

Zielke, G. (1984) A Survey of Generalized Matrix Inverses, *Comput. Math.*, 499–526.

Zielke, G. (1986) Report on Test Matrices for Generalized Inverses, *Computing*, **36**, 105–162.

Author index

Subject index

Mathematics and its Applications

Series Editor: G. M. BELL, Professor of Mathematics, King's College London (KQC), University of London

Numerical Analysis, Statistics and Operational Research

Editor: B. W. CONOLLY, Emeritus Professor of Mathematics (Operational Research), Queen Mary College, University of London